PIC Projects for
Non-Programmers

This book comes with a free version of Flowcode 4.5, which is available on its companion site at www.elsevierdirect.com/companions. This free version of Flowcode 4.5 has a few restrictions, which are removed if you chose to purchase a full license for the software, though this is not necessary in order to use the book.

For further information on Flowcode and about the author, please visit John Iovine's website at http://www.imagesco.com

PIC Projects for
Non-Programmers

John Iovine

AMSTERDAM • BOSTON • HEIDELBERG • LONDON • NEW YORK • OXFORD
PARIS • SAN DIEGO • SAN FRANCISCO • SINGAPORE • SYDNEY • TOKYO

Newnes Press is an imprint of Elsevier

Newnes is an imprint of Elsevier
The Boulevard, Langford Lane, Kidlington, Oxford, OX5 1GB
225 Wyman Street, Waltham, MA 02451, USA

First published 2012

British Library Cataloguing-in-Publication Data
A catalogue record for this book is available from the British Library

Library of Congress Number: 2011933374

ISBN: 978-1-85617-603-3

For information on all Newnes publications
visit our website at: www.elsevierdirect.com

Typeset by MPS Limited, a Macmillan Company, Chennai, India
www.macmillansolutions.com

Printed and bound by CPI Group (UK) Ltd, Croydon, CR0 4YY

CONTENTS

Chapter 9   Analog-to-Digital Converters

FLOWCODE PROGRAM INSTALLATION

This book comes with a free version of Flowcode 4.5, which is available on the companion site (please visit www.elsevierdirect. com/companions and follow the instructions there). The free version of Flowcode 4.5 has a few restrictions, which are removed when one purchases a license. The free version has a 2 KB code limit in addition to limited icon components. The student version of Flowcode 4.5, which as of 2011 is under $100.00 USD, eliminates these restrictions.

Minimum System Requirements

- Personal Computer
- Pentium processor or greater
- Windows 98, XP, Vista and WIN7 compatible
- CD Rom drive
- 256 MB RAM
- 50 MB hard disk space.

Step 1

Open Windows Explorer, copy the URL into the address bar and select the Flowcode software. Start the installation by running the "SETUP.EXE." The start-up installation screen is shown in Figure 1.1.

Step 2

If the install program asks you if you want to install the PPP program (see Figure 1.2), the PPP program is specific

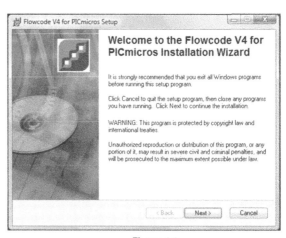

Figure 1.1

PIC Projects for Non-Programmers. DOI: 10.1016/B978-1-85617-603-3.00001-4

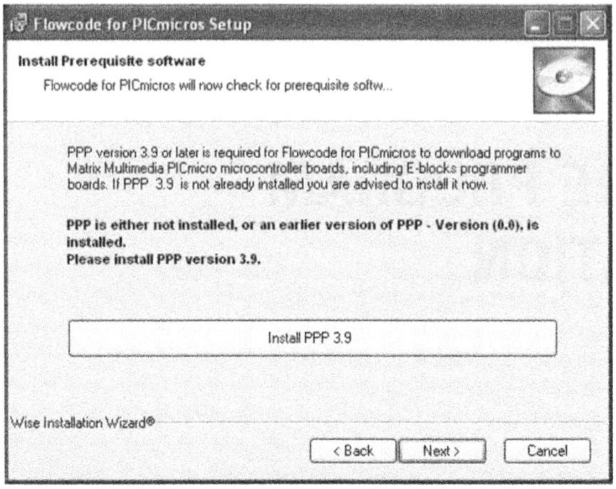

Figure 1.2

to the use of Matrix Multimedia's own PIC programmer hardware. If you elected not to use this programmer, you do not need this software. I chose to use the PPP programmer as well as microEngineering's EPIC USB programmer. The Matrix Multimedia PIC programmer is a combination programmer/ developer board that allows development. It has multiple DB9 connectors for connecting Matrix's proprietary I/O boards. This programmer is discussed further in Chapter 3.

Step 3

The next screen asks you to input your name and affiliate organization (if any) (see Figure 1.3). Fill in the required information and select "Next."

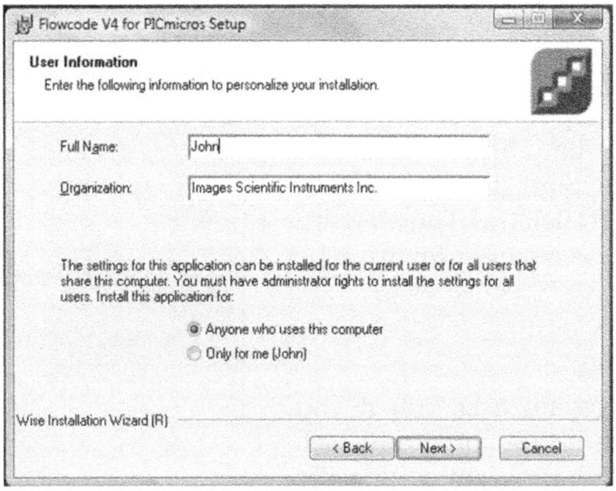

Figure 1.3

Step 4

The next screen is a standard license agreement form (see Figure 1.4). Select the acceptance radio box and hit "Next."

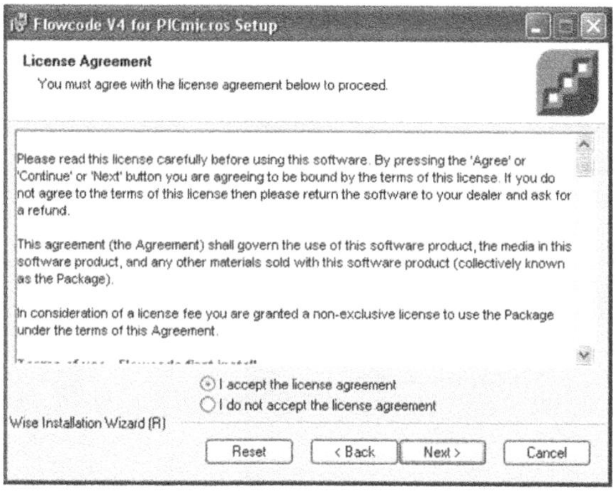

Figure 1.4

Step 5

Unless you have a reason to change the destination folder for the installation, keep the folder at its default location (see Figure 1.5) and select "Next."

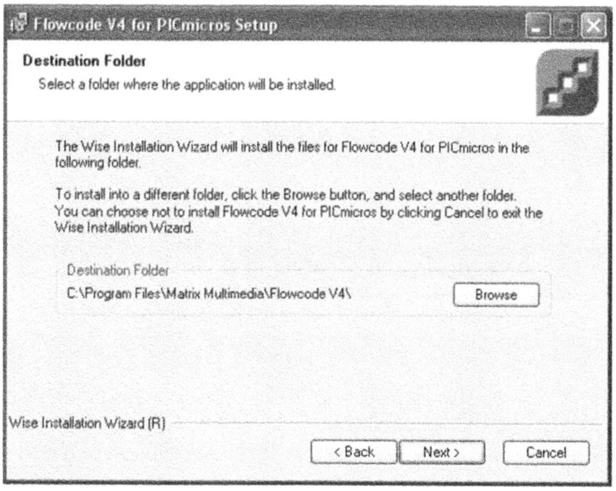

Figure 1.5

Step 6

The following install screen asks you to confirm features installation, followed by the default programmer (see Figures 1.6 and 1.7).

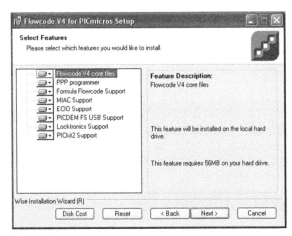

Figure 1.6

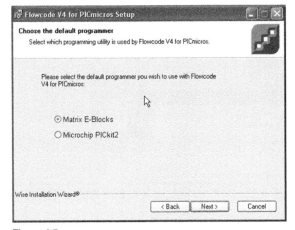

Figure 1.7

Step 7

The next screen asks you to confirm the installation, followed by the installation (see Figures 1.8 and 1.9).

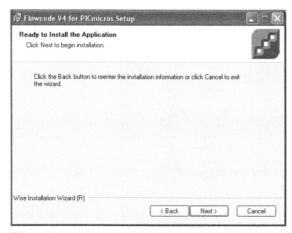

Figure 1.8

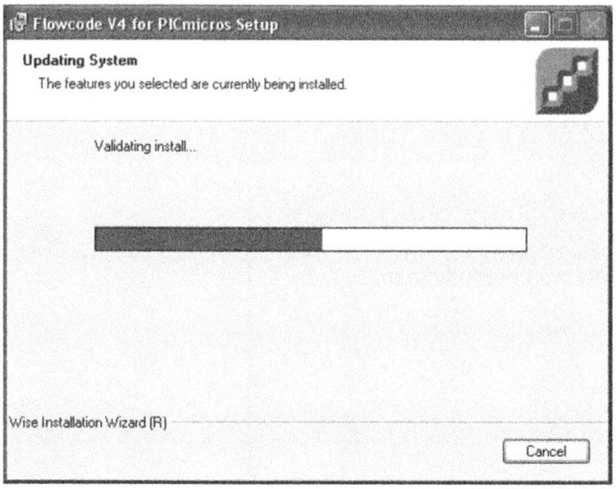

Figure 1.9

Step 8

The following screen asks for license key information (see Figure 1.10). If you are using the free version of Flowcode, select the "Do not enter a key" radio option and leave the key information blank. Hit the "Next" button.

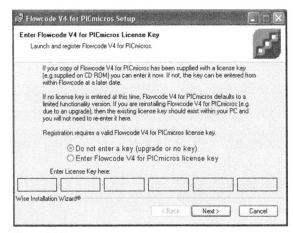

Figure 1.10

Step 9

The following screen informs and confirms that you installed Flowcode sucessfully (see Figure 1.11).

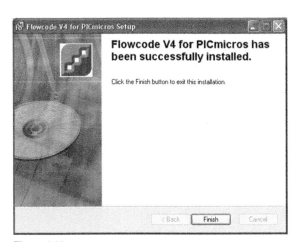

Figure 1.11

The following screen is the start-up screen for Flowcode (see Figure 1.12).

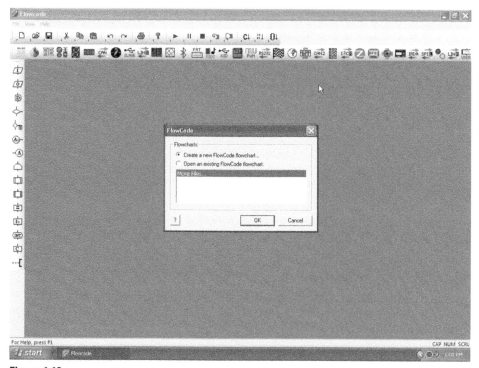

Figure 1.12

With Flowcode installed, proceed to Chapter 2.

2

WRITING OUR FIRST PROGRAM "WINK"

In Chapter 1 we installed Flowcode onto our computer. Now let's jump in and use it. Usually one explains the syntax (language) of a particular compiler before jumping in to write a program. Since Flowcode is a flow-charting graphic language technically there is no syntax, so let's see how easy it is to write a working program using its graphic interface. Being a graphic interface, there are screen images to guide you through.

Our first Flowcode program is titled "Wink." What Wink accomplishes is to alternately light two LEDs connected to two Input and Output (I/O) lines to a PIC microcontroller. While this is a simple program, we still have a number of options to consider before we start writing our flowchart program. Each option has its own pros and cons.

Option 1: Write the program and run the program within Flowcode's built-in simulator.

> **Pros**: Easy to do and immediate. Does not require PIC microcontroller, or any external components; no hardware is required. This is a great educational tool to get you up and running.

> **Cons**: Program execution is a simulation. It's not the real thing and things happen in the real world unaccounted for in simulations.

Option 2: Write the program; compile the program into Hex code. Upload the Hex code program (firmware[1]) file into a PIC microcontroller using a PIC-compatible programmer.

---

[1]**Firmware** is the programmable content of a hardware device, which in our case is the PIC microcontroller. Typically firmware is loaded into a hardware device and not updated. An example would be the various timing and wash cycles programmed into a washing machine or dishwasher's microcontroller.

PIC Projects for Non-Programmers. DOI: 10.1016/B978-1-85617-603-3.00002-6

Take the programmed microcontroller and prototype (build) a simple electrical circuit to test the program.

Pros: Real world implementation of software. This option provides experience building and testing of real world devices.

Cons: This option requires hardware and components, a PIC-compatible programmer and a spoonful of electronic components to build the test circuit.

Option 3: Write the program; compile the program into Hex code. Upload the program (firmware) directly into an ECIO28 or ECIO40 PIC microcontroller using the computer's USB port.

Pros: Same as those listed for Option 2, plus this option saves you from purchasing a standalone programmer. Instead it requires you to purchase a more expensive ECIO28 or ECIO40 USB-compatible microcontroller. The ECIO is a PIC microcontroller on a small carrier board containing a USB interface. It is used just like a PIC microcontroller to build a simple circuit to test the program. This option also requires a spoonful of electronic components to build the test circuit.

Cons: As mentioned, the ECIO is a more expensive alternative than a standard PIC microcontroller. If you were going to produce a product it would be far less expensive to buy a PIC-compatible programmer and program the microcontroller directly.

Which Option to Take?

I can't answer that question for you, so we will provide instructions for each option mentioned above. You decide your own path. The Flowcode program is the same for all the options, so the coding instructions will apply. However, there is an exception with option 3: when the Flowcode software (or new project file) is started you need to choose the ECIO28 or ECIO40 for the PIC microcontroller chip. In this chapter we will focus on the coding of our simple program and running the program in Flowcode's built-in simulator. Real world options 2 and 3 will be covered in Chapter 3.

Our first question is "Which PIC microcontroller should we choose to start writing code for?" There are a wide variety of PIC microcontrollers to choose from. I am inclined to use the PIC 16F88, and I will do so later on. To begin with, I want to start with the old 16F84. Why? I chose the 16F84 because this PIC chip has been around so long there are thousands of programs available on the Internet. The procedures you'll learn using this chip are directly applicable to later and larger PIC microcontrollers, so no learning is lost using the 16F84. However, the 16F84 is obsolete at this point in time. So if you prefer to begin your work using the 16F88 chip immediately, feel free to do so.

Selecting a microcontroller to write a Flowcode program isn't difficult; switching microcontrollers within Flowcode is as simple as choosing a microcontroller from a drop-down list of microcontrollers. So if after writing our 16F84 program, you wanted to use it for a 16F628 or a 16F88, simply choose that controller from the drop-down list and recompile the file. Or if you decide to start with the 16F88, simply select that microcontroller. We will discuss selecting microcontrollers later on in the book.

Writing the Flowchart

Start your Flowcode program. If you are working with the free version of Flowcode provided with this book, your opening screen resembles Figure 2.1.

Next you get to choose to open any existing files or to start a new Flowchart program. You choose to start a new program (see Figure 2.2).

Next you select the "target" microcontroller for your Flowchart program (see Figure 2.3). Select the 16F84 (or 16F88) from the list of available microcontrollers and hit OK.

Flowcode now opens up to its graphic interface (see Figure 2.4).

This screen needs a little explaining. When Flowcode opened it started a Flowcode chart for you titled "Main." It placed two icons in the flowchart, "Begin" and "End," since all Flowcode programs have a beginning and an end. Toward the right side of the screen Flowcode has generated a graphic of the selected target chip you choose when you started your program. This

Figure 2.1

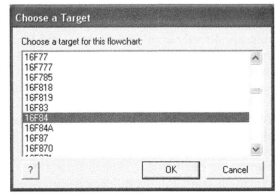

Figure 2.2   **Figure 2.3**

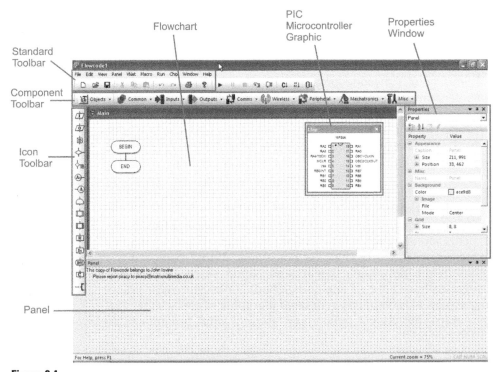

Figure 2.4

graphic displays the pin connections of the microcontroller and also provides visual indicators when the program is run in simulation. We'll get to that later.

Now let's look at the toolbars. We have a pretty standard Windows toolbar on top. Right underneath that we have our component toolbar. Each icon here represents a submenu of hardware components that you can add to your flowchart. For example: LEDs, switches, analog-to-digital converters, LCD modules. We'll cover most of these components in later chapters. Moving the mouse over a component icon opens up the submenu of components that can be selected. The ICON toolbar on the left side of the screen contains the available programming icons. The toolbars are dockable, so you can move and position them where you want them on your screen.

To use a programming icon, click on the icon and drag it onto the flowchart screen. When the icon is close to the flowchart, the pointer changes into a small yellow arrow indicating where the icon will be placed when the mouse button is released. The Properties window will show the properties of the element selected in Flowcode. The Panel space is where we put our program's components, such as LEDs.

We now have enough information to start coding. We want to add an output icon to our Flowcode chart. The output icon resembles the graphic on the left side of the screen in Figure 2.5. It is the second icon down from the top.

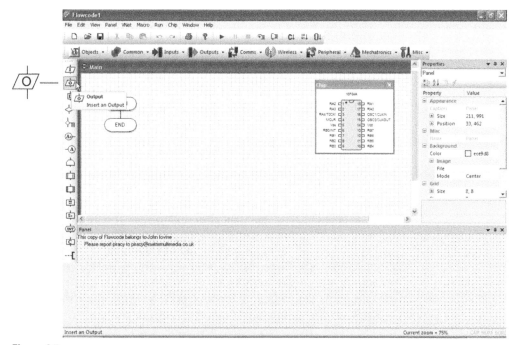

Figure 2.5

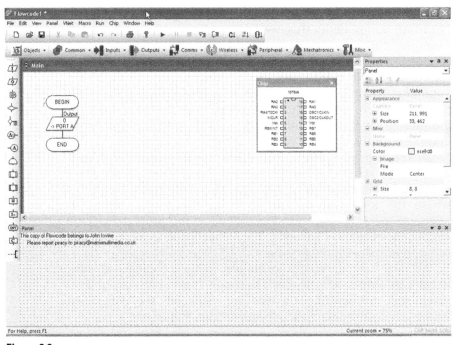

Figure 2.6

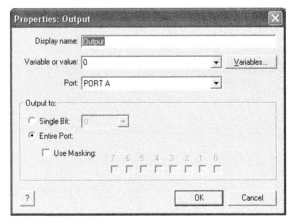

Figure 2.7

If you hold your mouse over an icon, the icon label appears. Click on the Output icon and drag it between the Begin and End icons and release the mouse button. Your screen should look like Figure 2.6.

The default properties of the Output icon are pointed to Port A of our micro-controller. We want to change that. Double click on the Output icon to open its Properties window (see Figure 2.7).

Put a value of 2 in the "Variable or value" box. Change the "Port": box to Port B; you can pull down the options arrow on the side of the box. The Properties box should look like Figure 2.8.

Click the "OK" button. We will now add a delay icon to the flowchart. The delay icon looks like the icon shown at the left of the screen in Figure 2.9. It is the third icon from the top.

Click on the Delay icon and drag it under the output icon and release. The delay icon has a default value of 1 millisecond. That's too fast. We need to increase the delay to 250 milliseconds.

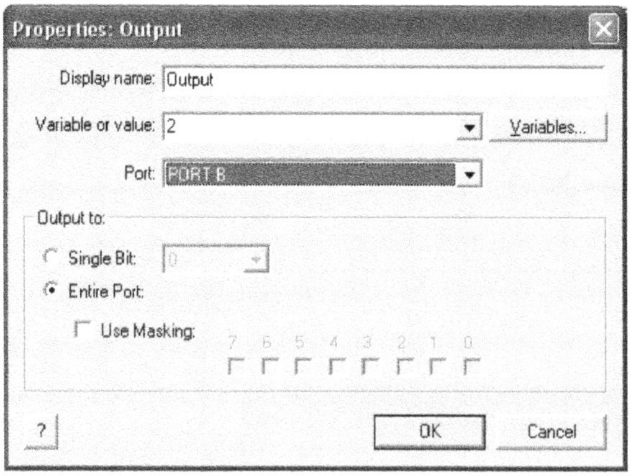

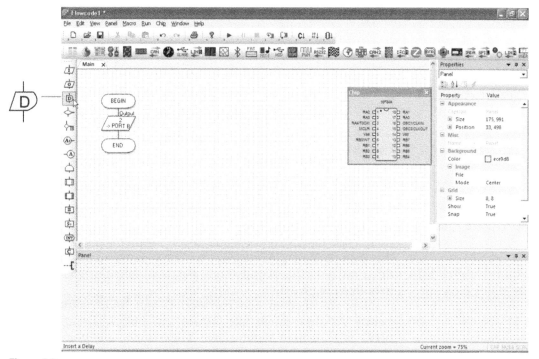

Figure 2.8

Figure 2.9

Double click on the delay icon on the flowchart to open up its Properties window (see Figure 2.10).

Change the delay value or variable to 250 and click OK.

We will now add two more icons like the last two. Click on the Output icon in the icon toolbar. Drag and drop the output

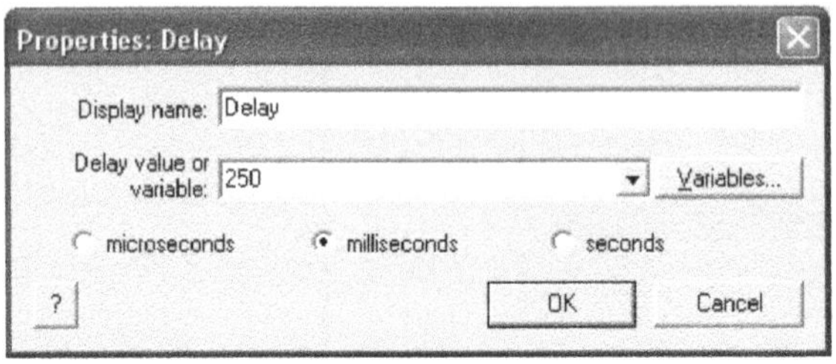

Figure 2.10

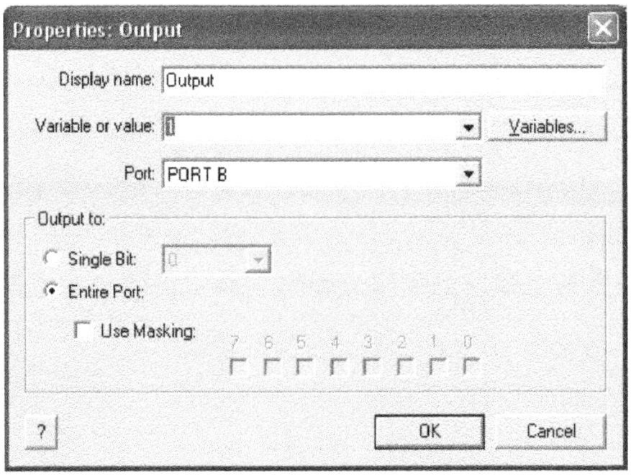

Figure 2.11

icon under the delay icon. Double click on the icon to open its Properties window, as shown in Figure 2.11.

Change the properties to Port B and the variable to 1 and then click OK. Next drag and drop another delay icon under this output icon. Double click on the delay icon to change the delay to 250 milliseconds and click OK. Your flowchart should look like Figure 2.12.

Now we need to add connection and jump points. These points are very much like "GoTo's" in basic. The connection point is shown in Figure 2.13. The connection point icon is shown on the left of the screen.

Drag and drop the connection point between the Begin and Output (2 Port B) icons. Your flowchart should look like Figure 2.14. In this figure I also closed the Properties box on the right side, grabbed the bottom edge of the flow chart area and moved it down a bit to provide a little more space.

Now we need to add our jump point. The jump point icon is shown in Figure 2.15.

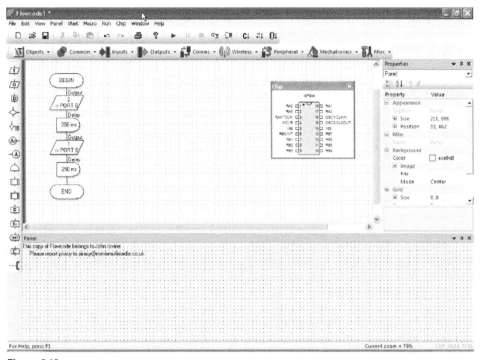

Figure 2.12

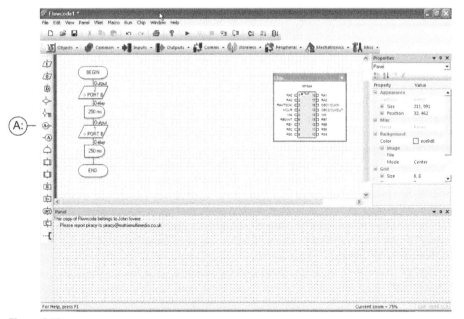

Figure 2.13

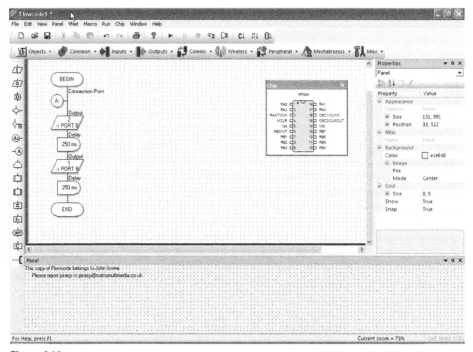

Figure 2.14

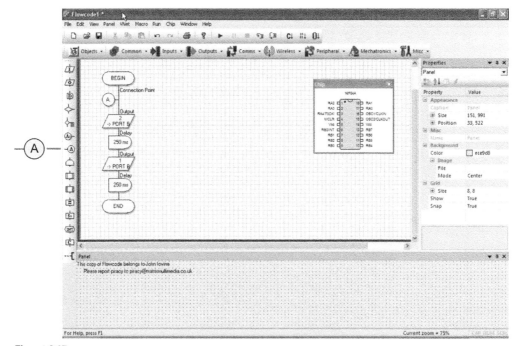

Figure 2.15

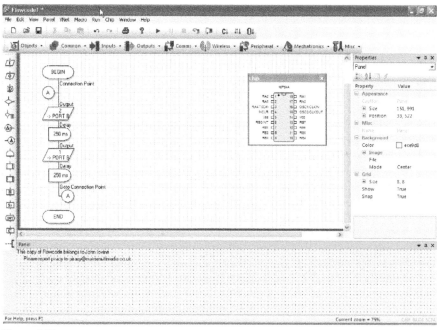

Figure 2.16

Drag and drop the jump point icon between the last Delay and End icon. So your flow chart now looks like Figure 2.16.

Technically your program is finished. This program is usable for all our three options. However, the one thing missing on the flowchart is a visual indication of the program operation. So in our next step we will add LEDs to the flowchart so we can have a visual indication that the program is running and that it is running properly or to our expectations. Later when we build the circuit in the real world, you will see the distinct difference between using LEDs in simulation and LEDs in the real world.

Option 1 Simulation

We are now going to use our component toolbar (Figure 2.17). The first component we will use is the LED array. The LED array icon exists under both the Common and Outputs menu items. In Figure 2.17 we are clicking on the LED array from the Outputs menu.

Open the Outputs menu and click once on the LED array icon. This puts an LED array in the panel portion of the Flowcode screen (see Figure 2.18).

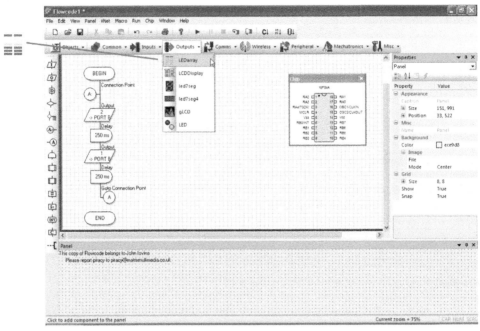

Figure 2.17

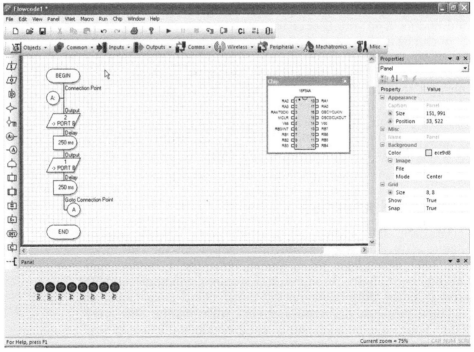

Figure 2.18

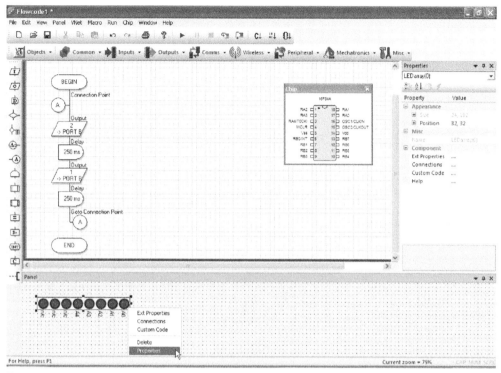

Figure 2.19

You can drag the LED array anywhere you like on the panel. We have to set the LED properties. If you had closed the properties window on the right hand side you need to reopen it now. Right click on the LED array and select "Properties"; this will open up the Properties window on the right (see Figure 2.19).

When you right-clicked on the LED array, you also had other options available (Ext properties, Connections and Custom code). These options are also available in the Properties window. To use the options in the Properties window make sure the LED(array0) is listed at the top of the Properties window; then, next to the property Connection, click on the "…" in the Value column. This opens up the dialog box shown in Figure 2.20.

Use the pull-down menu for Port, to change the designation from Port A to Port B (see Figure 2.21).

Now click "Done." In the Properties window (see Figure 2.19) now click on the "…" next to the "Edit properties" submenu item. A new dialog box opens as shown in Figure 2.22.

Figure 2.20

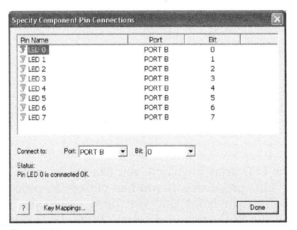

Figure 2.21

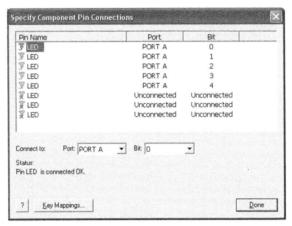

Figure 2.22

Change the number of LEDs to two. Change the LED color to green as shown in Figure 2.23 and click OK.

The change this dialog box produces is immediately obvious. We now have two green LEDs in our LED array connected to Port B. Your screen ought to look like Figure 2.24.

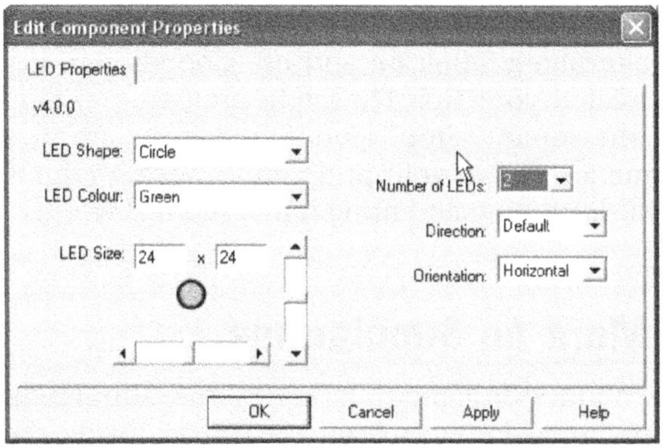

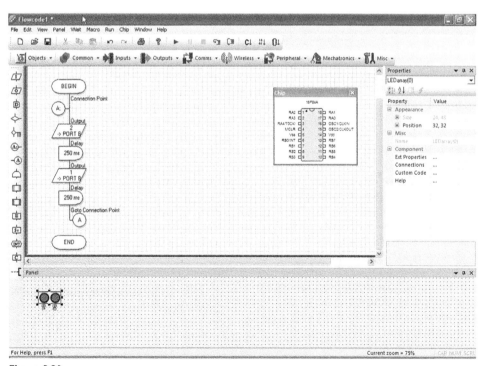

Figure 2.23

Figure 2.24

Figure 2.25

Running the Simulation

With our output components added to the screen we are ready to run our simulation to check the program function. Below the menu item "Help" on the standard tool bar we have three DVR type controls (see Figure 2.25).

To start the simulation, click on the arrow icon (see Figure 2.26).

The LEDs will alternately blink on and off. Congratulations! You have just completed your first Flowcode program. To stop the simulation hit the square "Stop" icon. Simulations are nice for checking the functionality of your program. Now that's done, we can take it out of simulation and bring it into the real world.

First a Little More on Simulations

But before we build our circuit, let's run the simulation again. Now look at the two I/O pins on the 16F84 inside the "PIC Microcontroller Graphic" window, to which we connected our LEDs. One LED is connected to Port B, bit 0 and the other to Port B, bit 1. Looking at our 16F84 graphic, RB0 pin 6 is Port

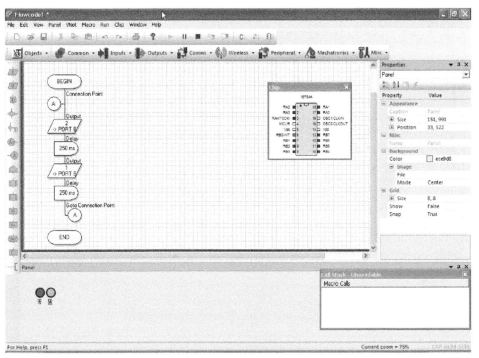

Figure 2.26

B bit 0 or PortB.0 and RB1 pin 7 is Port B bit 1 or PortB.1. When running the simulation, you ought to see each I/O pin alternatively turning red, signifying that the pin is high or at + 5 volts. Also the high pin is in sync with the lighted LED. Did your simulation show the I/O pins changing color? Sometimes this portion of the simulation doesn't always work, but we have an alternative method to show the I/O pin color change.

Step Mode(s)

On the top toolbar under the Run menu we have a few submenu options: Go continue, Step into and Step over. The two bottom menu options Pause and Stop are ghosted. These bottom menu options only become available when a simulation is running. Going back to the first of the three submenu options, Go continue starts the simulation, just like our VCR type control, or continues the simulation if you paused it. The Step into option allows you to step through the program. You can also use the menu option or use the F8 key to step through the program. Go back to the Flowchart window and hit the F8 key. A red rectangular box (which will appear red on your computer screen image) appears around the Begin icon (see Figure 2.27).

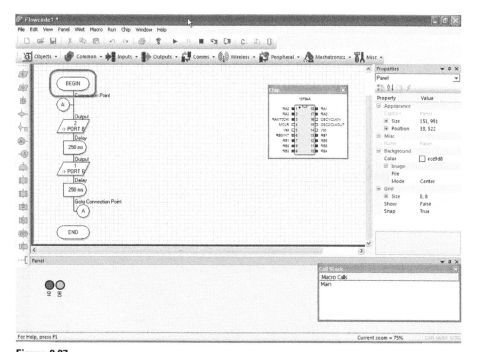

Figure 2.27

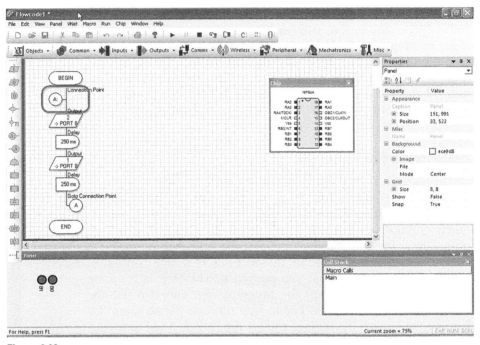

Figure 2.28

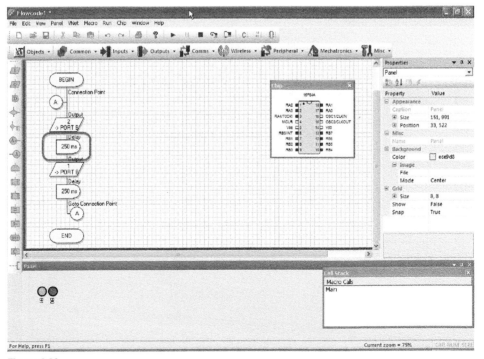

Figure 2.29

Each time you hit the F8 key the rectangular box appears around the next successive flowchart icon. In addition to surrounding the icon, whatever action is called for in the icon is carried out in the simulation (see Figure 2.28).

As you step through the program and bring the I/O pins high, the 16F84 graphic should illustrate the high pin by coloring the pin red on your computer screen, and of course the LED display functions as well, as shown in Figure 2.29.

Save Your File

At this point you ought to save your flowchart program. Go to File menu and select the Save submenu item. Title the program "Wink" and save it in a programs directory. Flowcode will automatically add the ".fcf" suffix to your program name (see Figure 2.30).

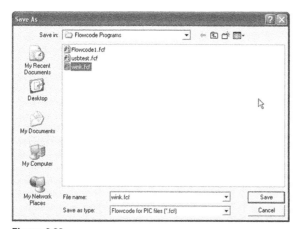

Figure 2.30

In the next chapter we will take our simple program and basic circuit and bring it out of the PC into the physical world.

3

UPLOADING "WINK" INTO MICROCONTROLLER(S)

In Chapter 2 we wrote a program in Flowcode and used Flowcode's built-in simulator to test the program. In this chapter we will build a physical circuit in the real word to test our program.

The first step to building the circuit is collecting the hardware and components. The solderless breadboard is essential hardware for building and prototyping temporary circuits. It allows you to quickly build circuits by plugging electronic components into the board and connecting them to other components without soldering. This makes it easy to change circuit wiring, switch out components and disconnect or reconnect components as you build and develop a circuit.

All the circuits used in this book have been built and tested. So if they don't work properly, look for wiring or component mistakes. We will be using a solderless breadboard extensively in our upcoming projects. Since it is an important component used throughout the book, I am going to explain its construction and use in detail so that you can start using it correctly from our first project onwards.

There are various sizes of solderless breadboards available (see Figure 3.1). What make a breadboard useful are the plug-in connections and the internal connection path. On the top of the board there are small plug-in holes approximately 0.100 inches apart. When a wire or component pin is plugged into a hole it makes electrical contact with a metal conductor strip underneath the plastic. Figure 3.2 shows a cutaway of a typical solderless breadboard and the underlying metal connector strips.

The blue-gray rectangles are the conductor strips that lie beneath the plastic cover. The two center strips run from the top of the board to the bottom. These strips are typically used

PIC Projects for Non-Programmers. DOI: 10.1016/B978-1-85617-603-3.00003-8

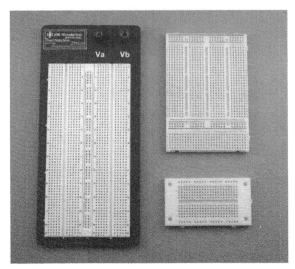

Figure 3.1

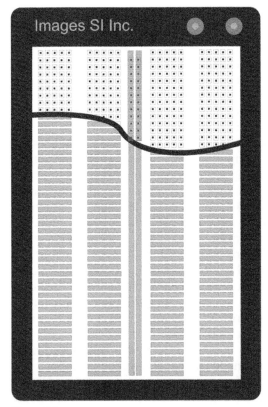

Figure 3.2

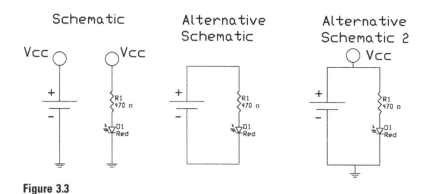

Figure 3.3

for power (the strip shaded in red for the +V and the strip shaded in blue for the ground). Let's build a simple electrical circuit on the solderless breadboard. Since the two power strips run the length of the solderless breadboard, it is easy to supply voltage or ground to any component placed on the board. The red and blue circles on the upper right of the breadboard represent screw terminals for connecting power.

Figure 3.3 shows a simple two-component LED circuit that is drawn three different ways. All we have in our circuit is an LED, current limiting resistor R1 and a 9 volt battery. When power is applied the LED will light up.

In the first schematic we have the negative leads of the battery and LED connected to ground. The positive leads are connected to a circle identified as Vcc. In the alternative schematic, we show the positive and negative terminals of our components connected by a wire. Both drawings are correct. In the second alternative schematic we identify which lines are Vcc and ground.

Electrical designers typically use the ground and positive terminal symbols to reduce the amount of lines in a schematic. It's not important in our simple drawing, but as schematics become more complex, ground and power lines can make the schematic more difficult to read; therefore it's easier to eliminate them from the drawing by using the symbols.

The electrical circuit applied to the solderless breadboard will look like Figure 3.4. The purpose of this exercise is to use the solderless breadboard, and become familiar with using it.

Before we move on, Figure 3.5 illustrates cutaways of two different solderless breadboards. They may also be used in the same manner.

With our simple exercise finished we can move on to building our real world circuit.

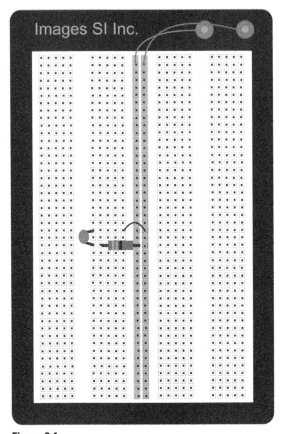

Figure 3.4

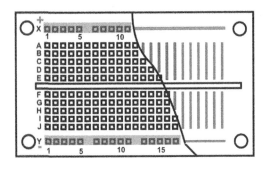

Figure 3.5

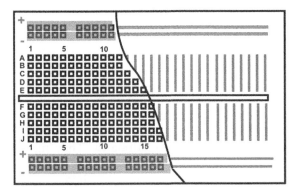

Real World Circuit (Option 2)

Unlike our simulation we need to add various real world components to make our microcontroller functional. The following is a parts list:

solderless breadboard
PIC 16F84 microcontroller
4 MHz Xtal
22 pF capacitors (2)
7805 voltage regulator
100 μF capacitor
300 μF capacitor
9 volt transistor battery cap
9 volt battery
4.7 K ohm ¼ watt resistor
470 ohm ¼ watt resistors (2)
subminiature LEDs (2)
on−off toggle switch
momentary contact switch N.O. (normally open).

Required Hardware

PIC-compatible programmer
Optional but highly recommended:
Inexpensive volt-ohm-meter (VOM) or digital multimeter.

The 16F84 (or 16F88) microcontroller expects +5 volts and a ground. To supply power to our microcontroller we use a 9 volt battery with a 7805 five-volt voltage regulator and two capacitors. The 7805 voltage regulator takes the 9 volts from the battery and outputs a regulated 5 volts.

The microcontroller needs a clock, something that will supply it with pulses. Each clock pulse increments its program instructions. Consider the clock the heartbeat of the microcontroller. There are a few clocks available for us to use; I choose to give our microcontroller a 4 MHz clock. The clock is composed of a 4 MHz crystal (Xtal) and two 22 pF capacitors. We need one more component to get our microcontroller moving: it's a 4700 ohm (4.7 K) resistor. The 4.7 K resistor connects to the +5 volt power supply and pin 4 (MCLR) of the microcontroller. The MCLR pin is a reset pin and is held high through the 4.7 K resistor. Held high means for normal microcontroller operation the pin is held at +5 volts.

If the program halts due to an unexpected error, we can reset the microcontroller. When we reset the microcontroller, the program is restarted from the beginning and runs. In many cases this is the preferred action to take rather than turning the entire circuit off and on again. To reset the microcontroller the MCLR pin (pin 4) is temporarily brought low. Brought low means it is temporarily connected to ground.

This is where the 4.7 K resistor plays an important part. For normal operation, the 4.7 K resistor keeps pin 4 close to +5 volts. When we connect pin 4 to ground to reset the microcontroller, the 4.7 K resistor limits the current between the +5 volts and ground. Without the 4.7 K resistor, a grounding pin 4 would make a connection between +5 volts and ground. This would cause a short circuit, and unimpeded current would flow, possibly causing damage to one or more components.

With the basic programs we will work through it is hard to imagine the need for a reset switch. As you gain experience writing and implementing programs, your programs will grow in size and complexity. There could be any number of reasons you would want to reset your microcontroller. The most common that I encountered is that the program gets stuck inside a loop and becomes non-responsive. A quick tap on a reset switch restarts the program from the beginning. Figure 3.6 shows a schematic of our circuit.

Figure 3.7 shows the circuit built on a solderless breadboard.

Before we apply power we must load our firmware (software/ program) into our microcontroller. To do this we must first

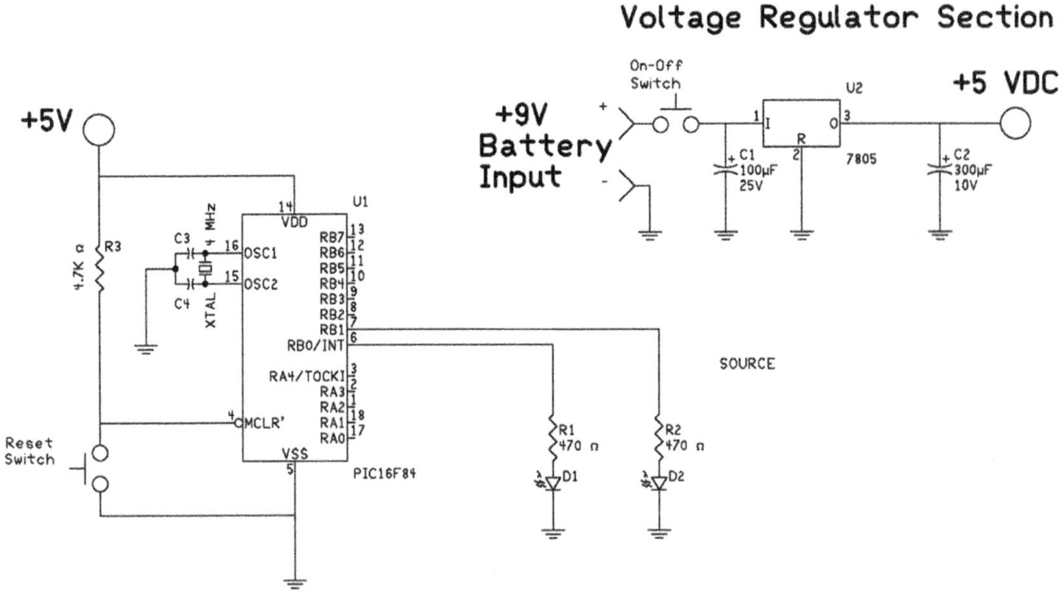

Figure 3.6

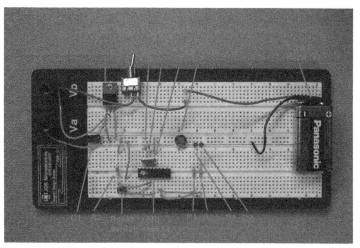

Figure 3.7

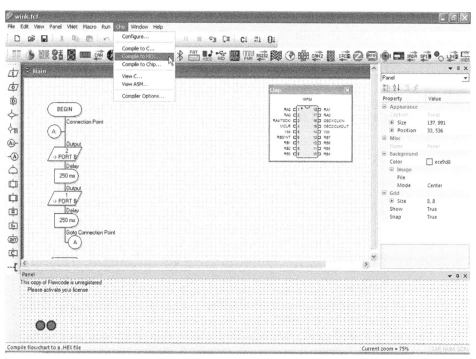

Figure 3.8

generate a Hex file using Flowcode. Start Flowcode and load the Wink program if it is not still open. Go to the top toolbar, select Chip menu item and then the Compile to HEX submenu item (see Figure 3.8).

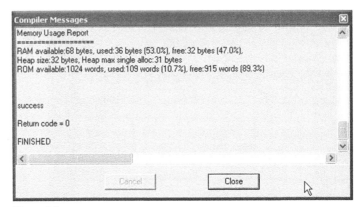

Figure 3.9

A compiler window opens up showing details the progress of the compilation. When it is finished, it will tell you the compilation was successful and is finished (see Figure 3.9). You can close the compiler window.

The Hex file is created in the same directory where you saved your original Flowcode file. To get the Hex file into the 16F84 PIC microcontroller requires the use of a programmer.

PIC Programmers

There are many PIC programmers on the market. Any programmer that is compatible with Microchip's standard Hex file should be able to load Flowcode's Hex files into the microcontroller. There are programmers that attach to the parallel port (printer port), serial port and USB port. I will only mention two programmers.

The Matrix Multimedia USB Programmer/Developer Board

I have used the Multiprogrammer from Matrix Multimedia (see Figure 3.10).

I have only a little experience working with Matrix's prefabricated application boards called "E-Blocks," and from my viewpoint these prefabricated boards are ideal for school and classroom applications, but not so much for the individual hobbyist learning microcontroller programming.

The reasons for my opinion follow; first is the expense of the E-Block modules. If you are teaching yourself programming, this is a

Figure 3.10

one-up proposition. So after you have learned and used your E-Block, your investment will most likely end up sitting in a drawer gathering dust. In addition, to utilize the E-Block properly, you need to purchase Matrix's PIC programmer which is a combination PIC programmer and development board, with ports connected to DB9 connectors for use with the E-Block system.

While I don't feel the system is applicable to the individual learning on his own, the situation changes dramatically for schools and classrooms. Imagine a microcontroller programming class that is filled with students jockeying to get their programming assignment completed within an allocated amount of class time. The prefab E-Block boards help by eliminating much, if not all, of the actual wiring and circuit building, and any errors associated with such building. For example, one of our first projects is wiring eight LEDs and resistors to a microcontroller. A prefabricated LED E-Block prewired with LEDs plugs right into the DB9 connector on the programming board (see Figure 3.11).

With the E-Block you know the wiring from the microcontroller to the LEDs is wired correctly. So if the program doesn't light the LEDs as expected, you immediately look toward the

Figure 3.11

program for the error and not the circuit. Therefore, in this classroom scenario the students remain focused on learning Flowcode programming.

In this book we use discrete components and handwire them into a solderless breadboard. Imagine the same class as described above filled with students building their microcontroller circuits on solderless breadboards. Now you need an instructor to look over the wiring before applying power to the circuit. It is tedious work and an instructor can easily miss an error when checking so many solderless breadboard circuits. If something malfunctions one has to check the circuit for wiring errors, as well as the program. For educational institutions, the E-Blocks are reusable from class to class, semester to semester and year to year. The initial investment in purchasing E-Blocks is amortized over the E-Block's lifetime, a worthwhile investment for classroom teaching.

So how does this affect you? As an individual, would you benefit from using the prefabricated E-Blocks? In my opinion, no, I don't think so. Primarily you are not under the same time constraints that exist in a classroom. To better understand using microcontrollers and applying them in circuits it helps to actually build the hardware. Even if the circuit fails, it forces you to troubleshoot the circuit (and/or program) until you find your error. This provides invaluable real world experience. You learn how everything fits together and works. You are wiring an LED to a port input/output pin on the microcontroller and using it in a program. Building on a solderless breadboard also forces you to build the supporting circuit for the microcontroller, including a regulated +5 volt power supply. In the real world there's not always a prefab to help you out, and when this occurs you'd be short on skills to create and build your circuit.

Since I focus on building circuits using discrete components on solderless breadboards, I decided on using a different programmer. The Matrix Multimedia USB programmer is a programmer/development board combination. Since I am not using E-Blocks in this book, I don't need the DB9 connectors or the development section of this PIC programmer. It is an excellent investment for school and classroom applications. If you decide to use this programmer, its software, PPPV3, is installed (or may be reinstalled) when installing the Flowcode software as described in Chapter 1. In addition, the Multiprogrammer's software and screens are similar to the software for the EPIC programmer that we discuss later on. Most of the instructions and information for using the EPIC programmer are also applicable to using this programmer.

EPIC Programmer from microEngineering

The programmer I use is microEngineering Lab's U2 Programmer (see Figure 3.12). I like this programmer because it is fast, versatile and completely powered by the USB port. The USB Programmer is shipped with a CD. Place the CD into your computer and run the set-up program.

It uses a ZIF (zero insertion force) adaptor socket to hold the microcontroller, which makes inserting and removing microcontrollers from the programmer simple and easy. The lever on the socket is moved to its vertical position to remove or insert a microcontroller. To lock the microcontroller down for programming, move the ZIF socket's lever to its horizontal position. Install the programmer's software according to the directions (see Figure 3.13), and we are ready to go.

After the program installs, attach the USB cable between your computer and the USB EPIC programmer and run the software (see Figure 3.14).

Figure 3.12

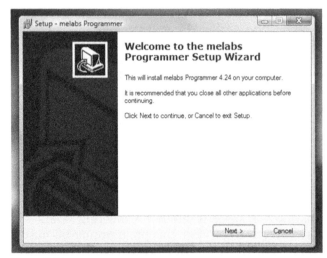

Figure 3.13

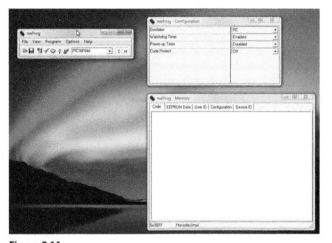

Figure 3.14

Use the File open menu and select the Wink.Hex Hex file (see Figure 3.15).

The program loads into the EPIC software (see Figure 3.16). You can see the hexadecimal code in the Code window. The window above the Code window is the Configuration window, and we need to make some changes here.

Next to the Oscillator, select the XT item in the pull-down menu. This changes the oscillator type from RC (resistor

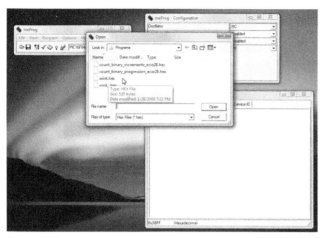

Figure 3.15

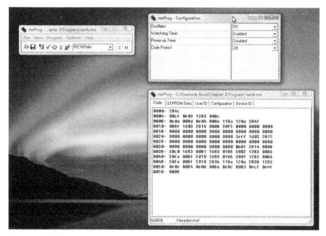

Figure 3.16

capacitor) to XT or crystal (XT). Change the Watchdog timer from Enabled to Disabled. Change the Power up timer from Disabled to Enabled. With the configuration changed, your screen should look like Figure 3.17. Next save the Hex file with the new configuration; this way if you need to reload the program you won't need to change the configuration again.

Place the microcontroller into the ZIF adaptor and select the Program submenu item from the Program menu. The program

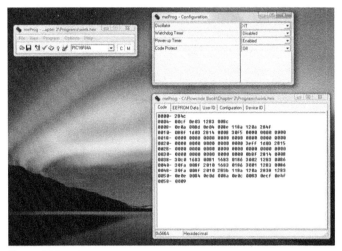

Figure 3.17

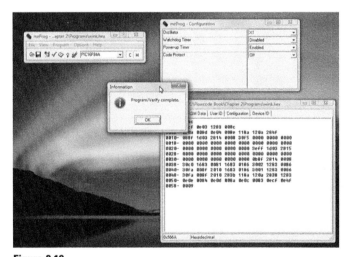

Figure 3.18

uploads the program into the microcontroller and gives you a Program/Verify window when complete (see Figure 3.18).

Take the microcontroller out of the ZIF adaptor and place it into your solderless breadboard. Apply power and you ought to see the LEDs turning on and off alternately, just as in your simulation (see Figure 3.7). If you have got this far and your circuit is working, congratulations! You have successfully programmed your microcontroller and got it running. If not, let's troubleshoot that circuit and find out why it's not running.

Troubleshooting the Circuit

Fortunately for us, this is a really simple circuit. Not too much can go wrong with it. However, if you need to trouble-shoot, this is when an inexpensive Multimeter or volt–ohm-meter (VOM) comes in handy (see Figure 3.19).

If you purchased an inexpensive multimeter, set it to read DC volts in the 9–10 volt range. First check the power to the circuit. You should have 9 volts going into the 7805 regulator and 5 volts coming out of the regulator. Then check the power connections to the microcontroller. Do you have 5 volts con-nected to pin 14 and pin 4? If "Yes," then check the LED diodes to make sure that they are orientated properly. If the LEDs are in backwards they will not light.

Change the multimeter to read ohms. When you touch the two leads together, you should read zero resistance or continu-ity. Now place one lead of the multimeter to ground (center ter-minal) of the 7805 voltage regulator and ground of the 16F84 microcontroller, pin 5. You should read zero resistance between these two points (continuity). Next check the ground of the microcontroller and ground to the two 22 pF capacitors. You can continue to check continuity between the leads of all the connected components.

One quick note: it is possible to configure Flowcode to auto-matically start the EPIC programmer from within Flowcode.

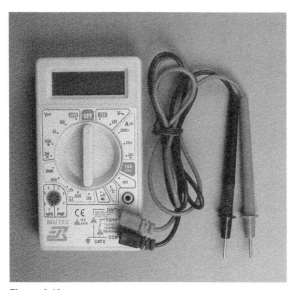

Figure 3.19

You would need to go to "Compiler options" under the Chip menu. From there, use the Browse window to select the EPIC programmer. I do not use this feature. I prefer to start the EPIC software and load the Hex file myself.

Real World Option 3

In this option we will use the ECIO28 microcontroller board (see Figure 3.20).

Using the ECIO28 allows us to program directly from the USB port of your computer without needing a programmer. So if you are just starting out, and don't know if programming microcontrollers is for you, you can pay a little extra money for the ECIO28 and put off the cost of purchasing a PIC programmer. The ECIO28 also simplifies construction and provides a few other conveniences. Using the ECIO28 reduces our component count for our simple circuit. The LEDs may be run directly from the ECIO and USB port. So the 9 volt battery, switch, battery cap, 7805 voltage regulator and its capacitors may be removed from the circuit.

Cost Versus Convenience

It's easy to fall in love with the ECIO28 and not drill down to the more primitive PIC microcontrollers. For schools offering programming courses and laboratory exercises, this is the ideal microcontroller to use. It would be hard to beat the convenience of the ECIO series of microcontrollers. This situation changes when working toward a commercial product or endeavor.

Figure 3.20

While basic microcontrollers are a little more inconvenient to set up and run, because of the support components when cost becomes a factor, you will not be able to match dollar for dollar performance with basic microcontrollers. Moving on, to bring our circuit into the real world using the ECIO28 we need the following parts list:

Solderless breadboard
ECIO28
330 ohm $\frac{1}{4}$ watt resistors (2)
subminiature LEDs (2).

As stated previously, for this project we can power our circuit directly from the USB port.

Before You Begin → ECIO Drivers

Your Windows operating system may not recognize the ECIO28 when it is first attached to the USB port. In that case jump over to Appendix A which shows you how to install the ECIO drivers. Once your drivers are installed, go to the Flowcode window. We need to change our selected target chip to the ECIO28. The target type can be changed by going to the View → Project options window. In here, select the ECIO28 or the ECIO40. The configuration for the ECIO devices is already set so you will not have to change the configuration. Compiling to the ECIO will also use the ECIO programming software. So under the View → Projects options window select ECIO28 as the target (see Figure 3.21) and hit OK.

The 16F84 screen graphic immediately changes to the ECIO28 graphic (see Figure 3.22).

The ECIO is shown as the 18F2455 PIC microcontroller. This microcontroller is the heart of the ECIO28. For fun, run the simulation. Notice the simulation runs identically.

Building the ECIO28 Circuit

This circuit is easier to build than the 16F84 because the power requirements are taken care of by the USB port. So for testing this simple circuit no external power is required. The schematic is shown in Figure 3.23.

Note that the pinout of the ECIO28 is not the same as the 18F2455 that is the PIC microcontroller used inside the ECIO module. This can get a little confusing since both the ECIO28 and the 18F2455 are labeled 18F2455. We plug the ECIO28 module into the solderless breadboard like a standard component

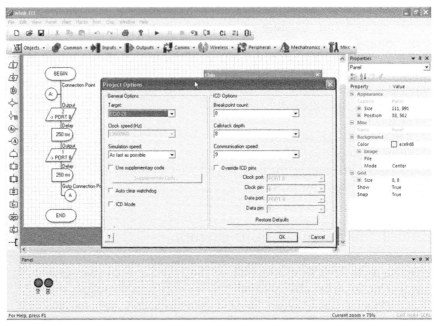

Figure 3.21

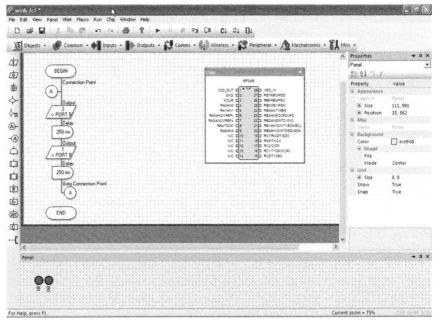

Figure 3.22

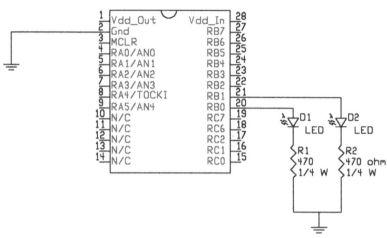

Figure 3.23

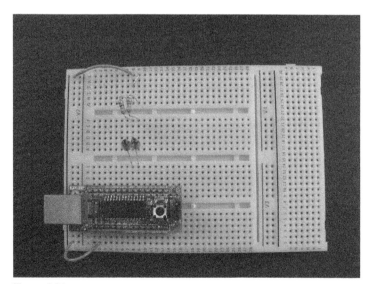

Figure 3.24

(see Figure 3.24). Run a USB cable from your computer to the ECIO28.

Choose from the main toolbar Chip; then Compile to chip (see Figure 3.25).

Follow the instructions on the screen. In my particular case I did not have the ECIO connected when I went to compile the

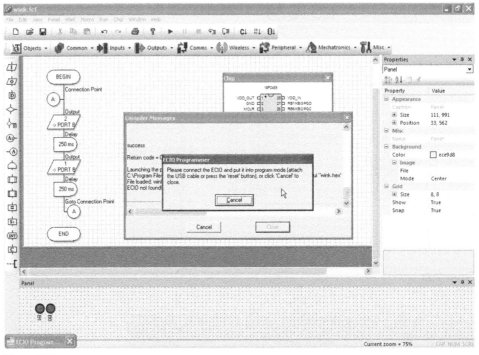

Figure 3.25

program. I simply connected the ECIO28 chip to the USB port and hit the reset button. The program automatically loaded onto the ECIO28 and began to run.

External Power

While our circuit can run off the USB port, when creating projects a USB cable connected to the project for power is inconvenient. So we can supply the ECIO28 and other components with external power (see Figure 3.26).

In this drawing an external 5 volt DC is connected to PIN 28 of the ECIO28. The jumper J4 that is positioned on a 3-pin header must be moved to the EXT side of the 3-pin header for external power to be applied to the ECIO28. For programming the ECIO28, the jumper must be positioned on the USB side of the 3-pin header. An external 5 volt power source can be built using the same components and drawing for the voltage regulator section of Figure 3.7.

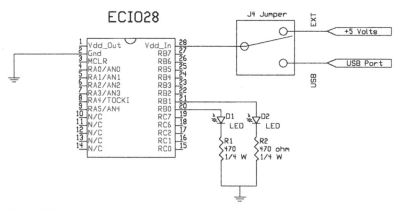

Figure 3.26

Preliminaries Finished

This concludes our preliminary work getting Flowcode up and running on your computer, and programming microcontrollers. In the following chapters the focus is on using Flowcode.

4

BINARY NUMBER SYSTEM — OUTPUTTING DATA

In this chapter I will introduce you to the binary number system. The binary number system is important because this is the number system which microcontrollers, as well as computers, use (and see). More than this, this is also how microcontrollers see the outside world and interact with it, as patterns of binary digits.

Bits

The smallest element in the binary number system is a bit. The term binary provides a hint to the values contained in a bit. Binary means two, as in 0 or 1, On−Off, Yes−No, True−False, etc.; you get the idea. A bit may contain one of two values. This number system is also called a Base 2 number system. The name bit is a contraction of the term **B**inary Dig**IT**. In computer lingo, a bit can have one of two values; 0 or 1. If dealing in a logic mode, the values are true and false where True = 1 and False = 0.

Base 2

Standard math is base 10, decimal. It is convenient for humans because it is based on having 10 fingers. Early computers, such as ENIAC, used base 10 arithmetic. However, it was quickly found that Base 2 binary worked much better for computers, so today all MCU and computers use binary. To represent large numbers, bits (binary digits) are assembled into groups. The most common groups are:

4 bits = nibble	decimal value 0−15
8 bits = byte	decimal value 0−255
16 bits = word	decimal value 0−65535
32 bits = long word	decimal value 0−4,294,967,295.

PIC Projects for Non-Programmers. DOI: 10.1016/B978-1-85617-603-3.00004-X

There are other groups of bits that include 7-bit for ASCII code, morse code, etc., that we are not going to look into except to note that they exist.

Bytes

A byte is a common digital number which we will be using often. It is composed of 8 bits. In binary the eight bits may be written as eight symbols, either a "1" or "0" as shown with these eight zeros 00000000.

8 bits	Decimal equivalent value
00000000	0
00000001	1

Table 4.1 Binary Progression

8 bits	Decimal equivalent value
00000000	0
00000001	1
00000010	2
00000100	4
00001000	8
00010000	16
00100000	32
01000000	64
10000000	128

Table 4.2

8 bits	Decimal	8 bits	Decimal
00000000	0	00001001	9
00000001	1	00001010	10
00000010	2	00001011	11
00000011	3	00001100	12
00000100	4	00001101	13
00000101	5	00001110	14
00000110	6	00001111	15
00000111	7	00010000	16
00001000	8	11111111	255

Bits are read from right to left. Table 4.1 below is a binary progression table; each progression of the "1" bit to the left doubles its value.

An 8-bit byte can contain any decimal value between 0 and 255. Table 4.2 shows the binary numbers 0 to 16. The progression continues until the maximum number for a byte is reached at 255.

Binary Addition

Rules for Binary Addition

Rule 1	$0+0=0$
Rule 2	$0+1=1$
Rule 3	$1+0=1$
Rule 4	$1+1=10$ (0 with a carry of 1 to next progression position)

Using these rules we can add binary numbers:

$(1+2=3)$	$(2+2=4)$	$(73+106=179)$	
00000001	00000010	01001001	Byte #1
+ 00000010	+ 00000010	01101010	Byte #2
00000011	00000100	10110011	Sum

How Microcontrollers Can Read and Write Binary Numbers

Microcontrollers do not have eyes or vision systems (yet) to enable them to see. So how do microcontrollers and computers read binary numbers? They see numbers as electrical values. A binary 1 will have an electric voltage of approximately +5 volts. A binary 0, on the other hand, will have a zero or ground voltage. When connecting to the outside world these electrical values are input (or output) on the microcontroller's pins that are assigned to Input and Output (I/O) functions.

So if we output a binary "1" to an output pin of the microcontroller it will output +5 volts on the pin. We can use that voltage to light an LED. The data specifications on the microcontroller will provide the maximum amount of current one can draw from any of its pins; usually it is a few milliamperes (mA). Typically one uses a resistor in series with the LED to make sure the LED doesn't draw more power than the I/O pin can safely supply. This is how the Wink program and the following counting programs light the LEDs connected to the microcontroller.

The inverse situation is also true. If we output a binary "0" to an output pin of the microcontroller it will place the pin at ground or zero volts. In this situation the pin can draw current in from a 5 volt power source connected to the pin. Again, the specification on the microcontroller will detail how much current an I/O can sink.

Eight-Bit Bytes and Ports

Eight-bit bytes are important for another reason. The Input/Output (I/O) lines (or pins) on PIC microcontrollers are arranged in 8-bit sized ports. The (8-bit) ports are identified as Port A, Port B, Port C, etc. The size and make of the microcontroller determine how many ports are available on any particular microcontroller. In addition, to save space, MicroChip, the manufacturer of the PIC series of microcontrollers, typically assigns multiple functions to a number of the microcontrollers I/O lines (or pins). The pin will have a default function and may be configured to other functions by setting or clearing bits in another 8-bit byte stored in the microcontroller's memory, called a register. We will deal with registers in later chapters. In the following binary counting programs we will use the 8-bit byte sized Port B to connect our LEDs.

Counting in Binary: Using LEDs to Represent Binary Numbers

We will modify our Wink program and expand it so it will use lit LEDs to count in binary. A lit LED is equal to a bit value of "1" and an unlit LED is equal to a bit value of "0." We need eight LEDs to represent each bit in one byte. The Flowcode

Table 4.2

8 bits	Decimal	8 bits	Decimal
00000000	0	00001001	9
00000001	1	00001010	10
00000010	2	00001011	11
00000011	3	00001100	12
00000100	4	00001101	13
00000101	5	00001110	14
00000110	6	00001111	15
00000111	7	00010000	16
00001000	8	11111111	255

program will count in binary from 0 to 255, as illustrated in Table 4.2 (which is repeated here for convenience).

The first thing to do is load your wink.fcf program into Flowcode. If you followed all the directions in the last chapter the chip selected is the ECIO28; let's switch back and use the 16F88 PIC microcontroller. The target type can be changed by going to the View → Project options window, as shown in Figure 4.1.

Select the 16F88 as the target chip as shown in Figure 4.2 and hit OK. If you want to you can continue using the 16F84 microcontroller. In later chapters we will be using the more advanced features the 16F88 offers, such as its built-in analog-to-digital converter.

Now save the file under a different name. Go to the menu option and perform a "save as" function to save the program under a new name such as "binary-counting.fcf." Select the LEDs shown in the bottom panel, then click on the three dots in the Properties window next to "Ext properties." This opens up a window as shown in Figure 4.3. Change the number of LEDs from 2 to 8, then hit OK.

Next, click on the three dots in the Properties window next to "Connections" as shown in Figure 4.4.

Then click on each LED under Pin name and change its Bit number to the next unused number, as shown in Figure 4.5.

When you are finished all LEDs will be connected to Port B bits 0 to 7 as shown in Figure 4.6; click on Done.

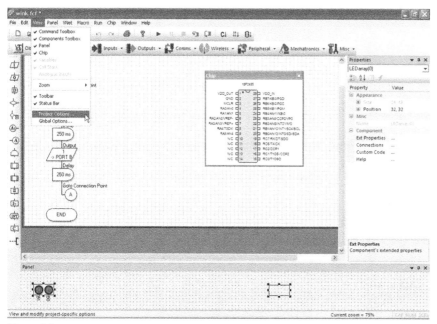

Figure 4.1

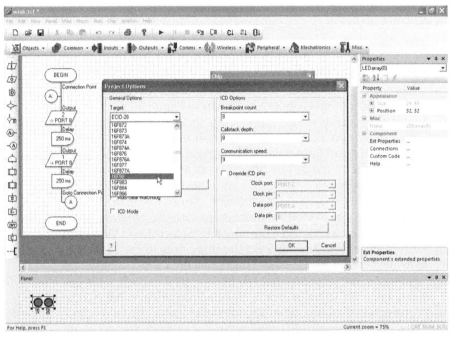

Figure 4.2

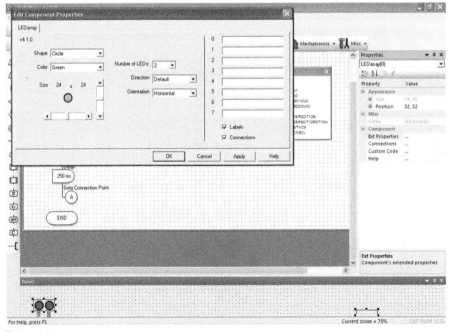

Figure 4.3

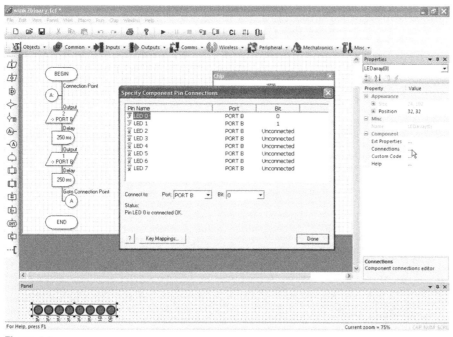

Figure 4.4

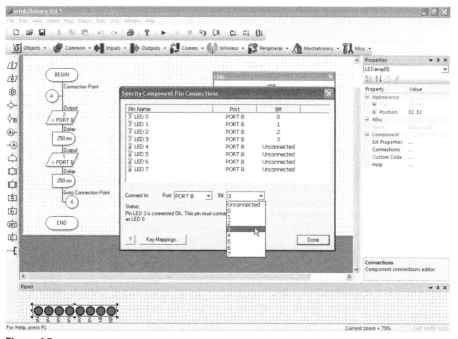

Figure 4.5

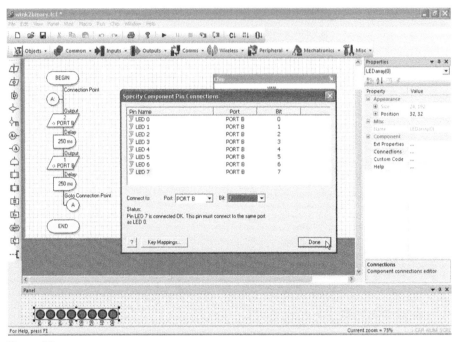

Figure 4.6

We now need to add a few more icons to the flowchart. Select the "Calculation" icon from the left toolbar as shown in Figure 4.7 and insert the icon between the "Begin" icon and the "A:" icon.

Once the icon is inserted, double click on the icon to bring up its Properties window, as shown in Figure 4.8.

We have no variable to work with in the calculation, so we must first create a variable. Click on the Variable button in the Properties window. That opens a Variable Manager window. In the Variable Manager window, click on the "Add New Variable" button. This opens a "Create New Variable" window. In the Create New Variable window, type the letter "X" in the "Name of new variable" text box. Select "Byte" under the variable options as shown in Figure 4.9.

Then click OK. The variable is now created. Close the Variable Manager window. The Properties Calculation: window should still be open. Type in "X = 0" as shown in Figure 4.10 and click OK.

Select the second Output icon on the flowchart, as shown in Figure 4.11, and hit the Delete key, to delete the icon.

Then select the second Delay icon and delete that also. Select the Calculation icon from the left toolbar and place

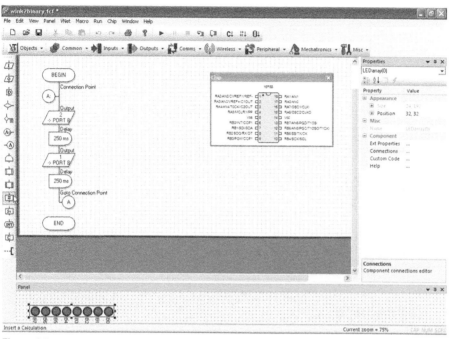

Figure 4.7

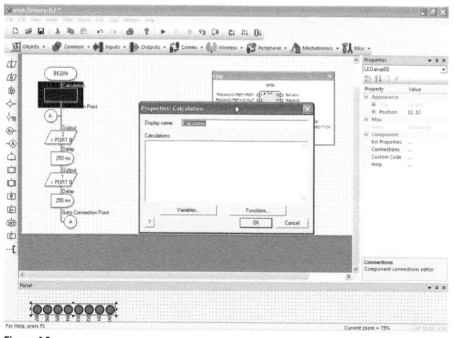

Figure 4.8

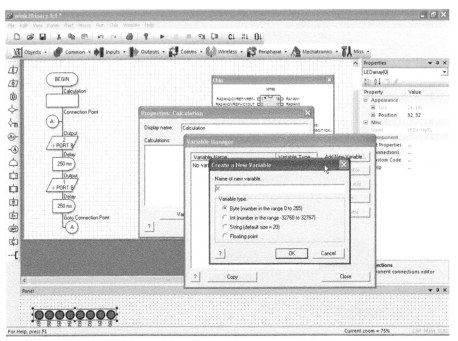

Figure 4.9

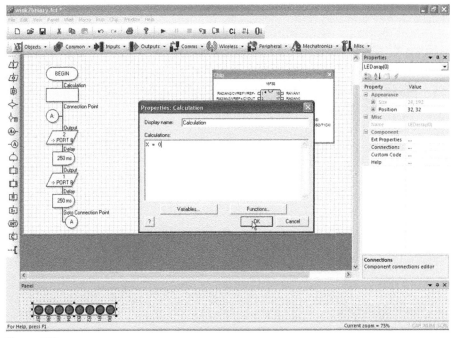

Figure 4.10

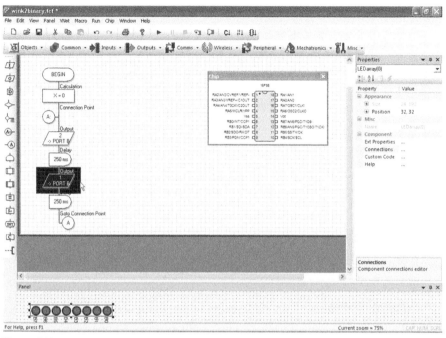

Figure 4.11

another Calculation icon between the Delay icon and the "A" icon. Double click on the icon to bring up its Properties window and type in "x = x+1" in its calculation text box, as shown in Figure 4.12, and hit OK.

Next double click on the Output icon "2 → PortB"; this will open up its Properties window. In the Variable or value text box either select the "X" variable using the pull-down menu or type in "X," as shown in Figure 4.13, and hit OK.

At this point you can hit the control to run the simulation. The program will begin counting and displaying the count using the eight LEDs. When the count reaches 255, the maximum number a byte can hold, it resets back to zero and begins counting again. It takes a little over a minute for the program to count to 255. If you want the program to count faster, double click on the Delay icon and reduce the delay time to 100 or 50 ms.

Before we move on to hardware, let's look at our flowcharts. Figure 4.14 is a flowchart of our starting program Wink.

Figure 4.15 is a flowchart of our Counting program.

The procedures for adding, deleting and opening up icon properties are the same for all icons. So as we proceed further,

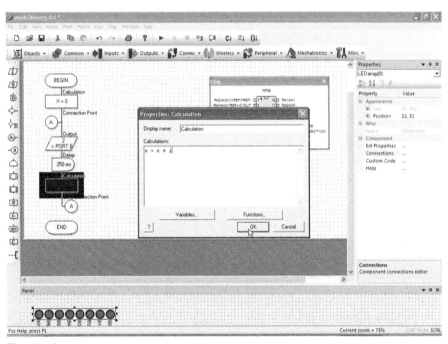

Figure 4.12

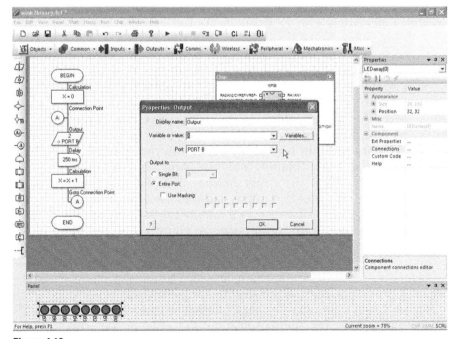

Figure 4.13

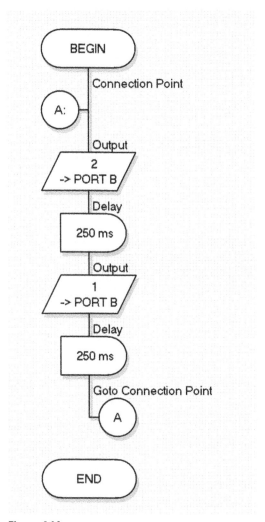

Figure 4.14

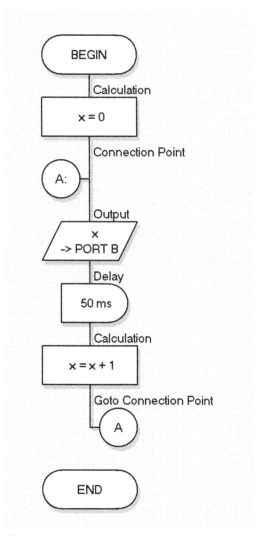

Figure 4.15

I will use finished flowcharts more often and point out new icons and property values as we work through the new programs. Before we move to the hardware, compile your program to a Hex file. Go to the top toolbar, select Chip menu item and then the Compile to Hex submenu item (see Figure 4.16).

When the compiler is finished, close the Compiler message window (see Figure 4.17).

Your program is compiled into a Hex file (firmware) and is ready to be uploaded into the microcontroller.

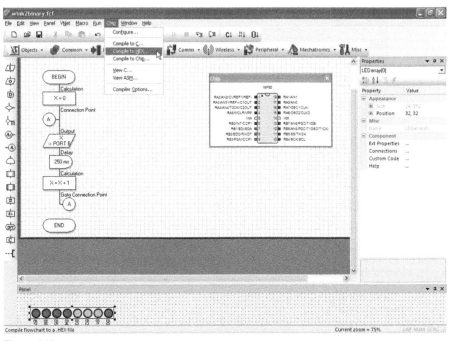

Figure 4.16

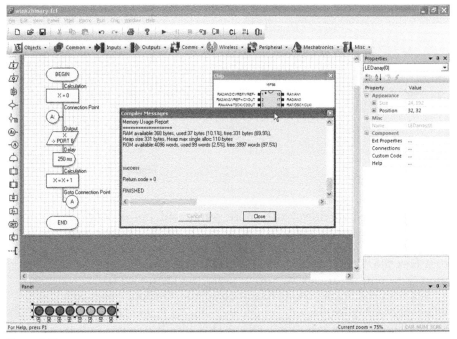

Figure 4.17

Hardware

In the hardware, we need to connect six more LEDs to make a total of eight LEDs connected to our microcontroller circuit. In this schematic I changed the PIC microcontroller from the 16F84 to the more updated 16F88 microcontroller. Essentially we are adding six more LEDs and resistors (see Figure 4.18).

Programming the 16F88 PIC Microcontroller

If you are using the EPIC programmer I recommended, when it's time to upload your firmware (Hex file program) into the microcontroller you need to change a few Configuration settings.

Run your EPIC software. Locate the Hex file and load it into the EPIC software. Look at Figure 4.19. This shows three of the EPIC windows overlaid on top of the Flowcode program. If the three windows are not showing when you start your EPIC software, go to the View menu item on the top toolbar, and select Configuration and Memory. You don't need to have the Memory window open. That's my personal preference; I like seeing the memory content of my Hex program. The memory window shows the program data in hexadecimal format; we'll get to what "hexadecimal" is later.

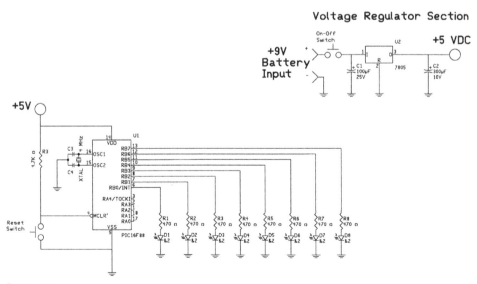

Figure 4.18

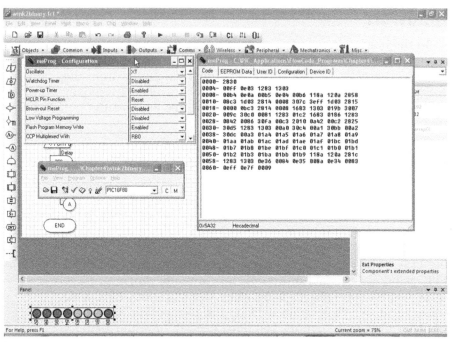

Figure 4.19

We came to make changes in the Configuration, and unless you make the following changes in the Configuration window, the program will not run properly when installed into the microcontroller. Use the pull-down menus on the left to change the configuration settings. Starting from the top:

Item	Set to:
Oscillator	XT
Watchdog timer	Disable
Power-up timer	Enable
MCLR pin	Reset
Brown-out reset	Disable
Low voltage programming	Disable
Flash programming memory write	Enable
CCP multiplexed with	RB0

Once these changes are made, perform a File → Save from the Epic software to save the configuration changes to the Hex file.

Place the microcontroller into the Epic ZIF adaptor, and select the Program submenu item from the Program menu. The program uploads the Hex file program into the microcontroller and gives you a Program/Verify window when complete. Place your microcontroller into the circuit built on the solderless breadboard and apply power (see Figure 4.20).

Figure 4.21 shows the schematic for the ECIO28.

To program the ECIO28, follow the same procedure that we used in Chapter 3. Go into the View → Projects options window and select ECIO28 as the target. Make sure the power jumper is set to USB or you will not be able to upload the new program. Run a USB cable from your computer to the ECIO28. When you connect the ECIO28, the previous ECIO28 program Wink will start running. That's OK; let it run. Choose from the main toolbar Chip, then Compile to chip, as we did in Chapter 3. Flowcode will ask you to save the "new" program with the ECIO28 targeted

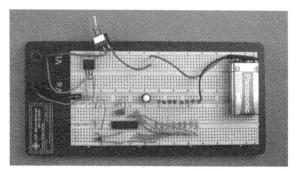

Figure 4.20

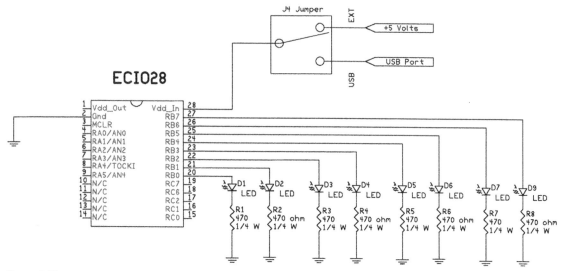

Figure 4.21

chip; save the file to allow the compiler to run. Flowcode will then open a "Compiler messages" window, compile the program and relay a message to you stating that the ECIO is not connected. Simply press the reset button on the ECIO28 to allow Flowcode to see the ECIO28, upload the firmware to it and start running the program (see Figure 4.22).

Figure 4.23 is a photograph of the ECIO28 circuit.

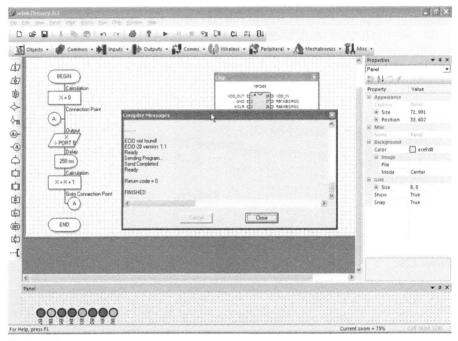

Figure 4.22

Figure 4.23

To fit everything on one board as shown in the photograph, I modified the LEDs by soldering a 330 ohm resistor to the positive leg of the LED, as shown in Figure 4.24.

Binary Progression

The previous programs counted in binary using lighted LEDs. Now let's light the LEDs in a linear progression as shown in Table 4.1. The binary progression table shows each progression of the "1" bit to the left doubles its numerical value.

Base 10

Figure 4.24

Base 10 has ten symbols: 0, 1, 2, 3, 4, 5, 6, 7, 8 and 9. When we need to express a number greater than 9 we use two or more digits. Each progression of a digit to the left increases its numerical value by the power of 10. To implement our new program we will use the flowchart. Figure 4.25 shows our binary counting program and our new binary progression program.

The first thing to notice is that we are using a different variable in the new program. Technically we don't need a new variable; we could still use X, but, to use X, we have to either change its variable type from being a Byte to an Integer, or create a program that uses a byte value, but by doing so the program may not appear to be so straightforward to everyone. So to keep things simple, we will create a new variable and after we have the programs running properly we'll look back to using the X variable.

Creating the Variable

Double click on the first calculation icon that contains X = 0. This opens the "Calculation: Properties" window. Click on the Variables button. This opens the Variable Manager window where you can add a new variable. Click on the "Add new variable" button. The "Create New Variable" window opens. Type the letter "Y" in the "Name of new variable" text box. Then select "Int" as the variable type, and hit OK (see Figure 4.26).

Next close the Variable Manager window. In the Properties: Calculation window type in "Y = 1" and hit OK (see Figure 4.27).

Now we need to change the properties of a few icons. In the "Output" icon, double click on it and change the variable from X to Y in the "Properties: Output" window that opens, then hit OK. Change the delay from 50 ms to 100 ms. In the second calculation icon change "X = X+1" to "Y = Y * 2". Next we need to add a decision icon (see Figure 4.28).

Drag the Decision icon between the Calculation icon and the Goto connection icon and release. Double click on the Decision

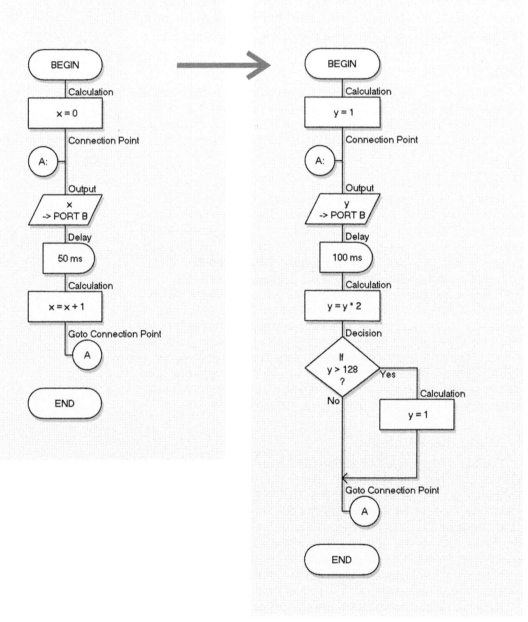

Figure 4.25

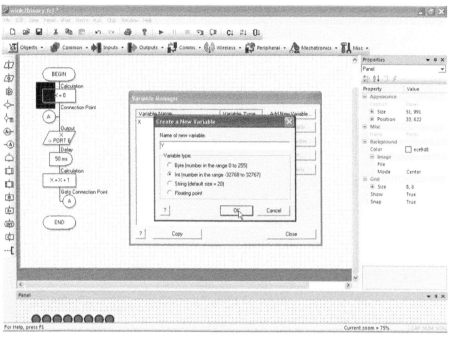

Figure 4.26

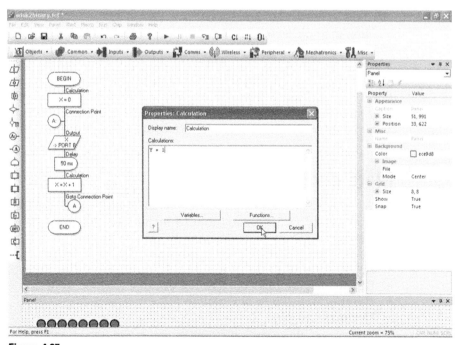

Figure 4.27

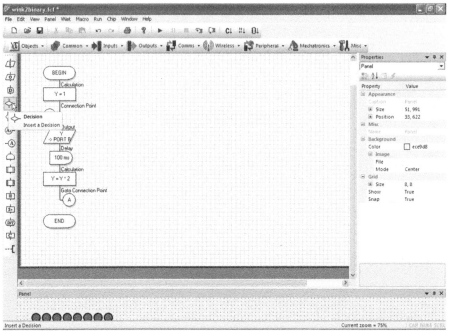

Figure 4.28

icon to open up the "Decision: Properties" window, type in the "If" text box Y > 128, and hit OK (see Figure 4.29).

Next grab a Calculation icon and put it into the "Yes" branch of the Decision icon. Double click on the new Calculation icon and type in "Y = 1" in the text box (see Figure 4.30), and hit OK.

The program is finished. Hit the Run icon to run the simulation. The hardware we have built requires no change to run this new program. Just program your PIC microcontroller or ECIO28 as shown previously with the new program and you are ready to run.

More on Binary

We looked at tables that related binary numbers to decimal numbers. Table 4.3 expands that to include hexadecimal (Hex) numbers and binary code decimal (BCD).

We can write a byte in four notations:
- Binary (Bits) %00000000 to %11111111;
- BCD which are bits coded to the decimal number systems;
- Hexadecimal $00 to $FF (00h to FFh);
- Decimal 0 to 255.

Binary we discussed previously, so I will not repeat that information.

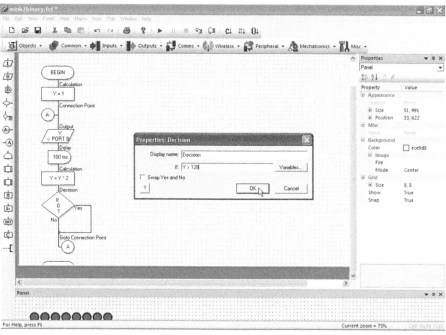

Figure 4.29

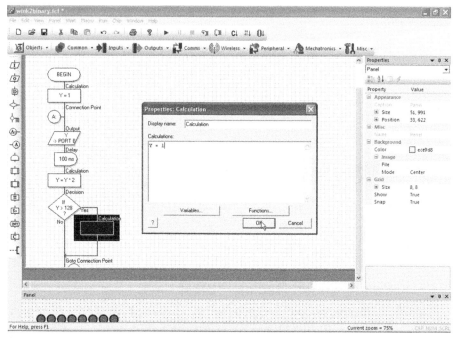

Figure 4.30

Table 4.3

Decimal		Binary	Hex		BCD
0		0000	0		0000
1		0001	1		0001
2		0010	2		0010
3		0011	3		0011
4		0100	4		0100
5		0101	5		0101
6		0110	6		0110
7		0111	7		0111
8		1000	8		1000
9		1001	9		1001
10		1010	A	0001	0000
11		1011	B	0001	0001
12		1100	C	0001	0010
13		1101	D	0001	0011
14		1110	E	0001	0100
15		1111	F	0001	0101
16	0001	0000	10	0001	0110
17	0001	0001	11	0001	0111
18	0001	0010	12	0001	1000
19	0001	0011	13	0001	1001
20	0001	0100	14	0010	0000
21	0001	0101	15	0010	0001
22	0001	0110	16	0010	0010
23	0001	0111	17	0010	0011
24	0001	1000	18	0010	0100
25	0001	1001	19	0010	0101
26	0001	1010	1A	0010	0110
27	0001	1011	1B	0010	0111
28	0001	1100	1C	0010	1000
29	0001	1101	1D	0010	1001
30	0001	1110	1E	0011	0000
31	0001	1111	1F	0011	0001
32	0010	0000	20	0011	0010

Binary Coded Decimal (BCD)

Binary coded decimals were designed to convert binary numbers to decimals. Typically they are used in electronic displays for clocks, VOM multimeters, thermometers and panel displays. The BCD digits increment when the count reaches 10 instead of 16 as is the case with standard binary.

Hexadecimal Numbers (Hex)

Hexadecimal numbers are base 16 numbers used extensively in digital electronics and microcontrollers. The base 16 is perfect for 4-bit nibbles. Since a byte is 8-bits, two nibbles can represent a byte, and therefore two Hex numbers may also represent a byte.

Programming Challenge

Using what you learned in this chapter, see if you can design a new binary progression program that reverses direction as it reaches each end of the LEDs (answer in Appendix A).

5

READING DATA ON INPUT LINES

In Chapter 4 we focused on outputting data (binary numbers) to an 8-bit I/O Port B. We connected LEDs to Port B pins to display the data (binary numbers) being output on the microcontroller port. In this chapter we will use a port to input data. We will read those data and make decisions based upon the data.

Electrical Signals

As stated in previous chapters, microcontrollers see a binary "1" as approximately + 5 volts and a binary "0" as approximately 0 volts (or ground). We want to supply binary "1"s and "0"s to an I/O line of the microcontroller and have the microcontroller read its value. Figure 5.1 shows two basic switch configurations that may be connected to an I/O pin of a microcontroller. The switch label SW1 outputs a normally high signal – binary "1" – to an I/O pin. When the momentary contact button is pressed, the output of the configuration switches to a binary low signal.

The switch labeled SW2 operates in an opposite manner to SW1. This information lets us add a switch to our LED display

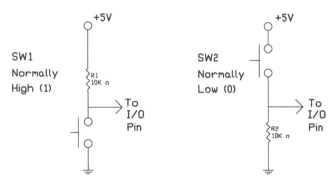

Figure 5.1

PIC Projects for Non-Programmers. DOI: 10.1016/B978-1-85617-603-3.00005-1

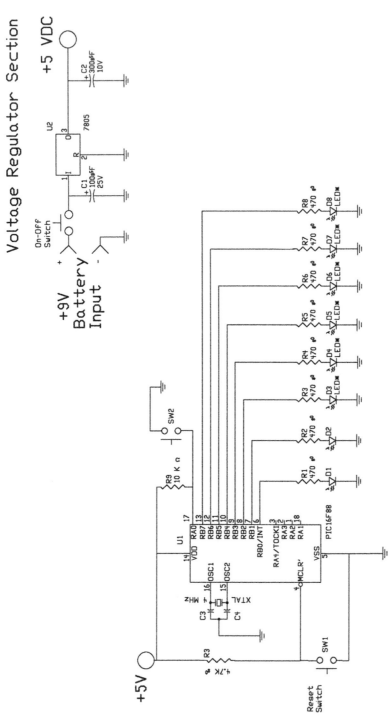

Figure 5.2

circuit. We will use the circuit to stop and start the binary counting of the program. Figure 5.2 shows the inclusion of a switch connected to pin 17, labeled RA0. This is a normally high switch configuration.

Figure 5.3 is the new flowchart to implement the switch function.

To implement this flowchart we need to add one component (switch), two icons (Input and Decision) and a new variable Y.

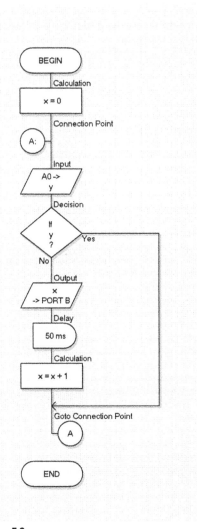

Figure 5.3

The Input Icon First

a. Select the Input icon from the left toolbar and drag it under the A: connection point (see Figure 5.4).

b. Double click on the Input icon to bring up its Properties window.

c. Change Input to "Single bit."

d. Click on "Variables" to bring up the Variable Properties window.

e. In the Variable Properties window create an integer variable labeled "Y."

f. Close the Variable Properties window. This drops you back down to the Input Properties window.

g. In the Input Properties window, select the variable "Y" in the Variable selection box (see Figure 5.5).

h. Hit OK.

Next We Add the Decision Icon

a. Select and drag the Decision icon under the Input icon and release.

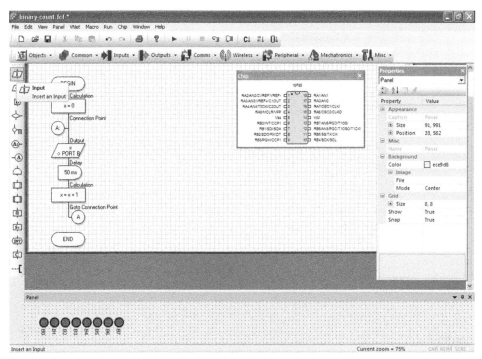

Figure 5.4

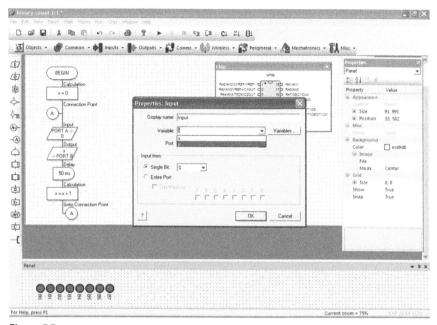

Figure 5.5

b. Double click on the Decision icon to bring up its Properties window.

c. In the If: text box, type in and replace the "0" with a "Y" (see Figure 5.6) and hit OK.

Next we need to move our icons into the Decision icon branch. Select the Output icon with your mouse. Keeping the left mouse button pressed, drag over the small arrow (which represents the Output icon) to the outside branch of the Decision icon (see Figure 5.7).

Then release the mouse button. The Output icon will have been moved to the outside branch of the Decision icon (see Figure 5.8).

Using the same procedure, select and drag the Delay icon and Calculation icon to the outside branch of the Decision icon. When you are finished your screen will look like Figure 5.9.

We are almost ready to run. To finish the program we need to add a switch. Go to the Common menu item on the top toolbar and select the Toggle switch (see Figure 5.10).

Hit the left mouse button with the Toggle switch highlighted; this places a Toggle switch in the bottom panel. The Toggle switch will be placed in the upper left corner; grab the switch and drag it to the center of the panel (see Figure 5.11).

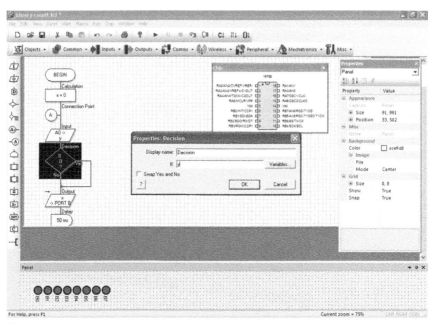

Figure 5.6

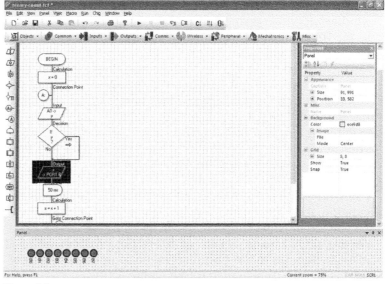

Figure 5.7

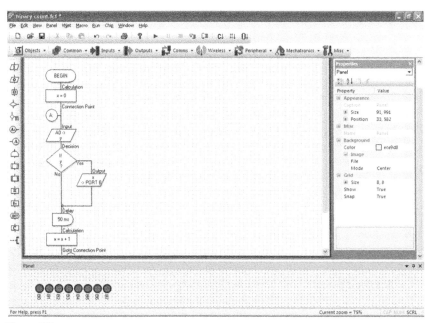

Figure 5.8

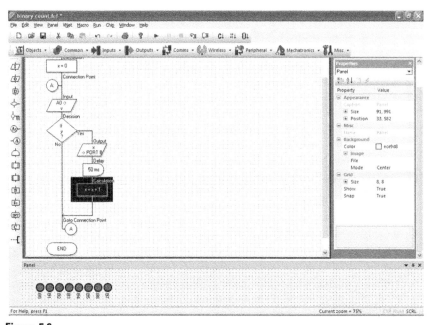

Figure 5.9

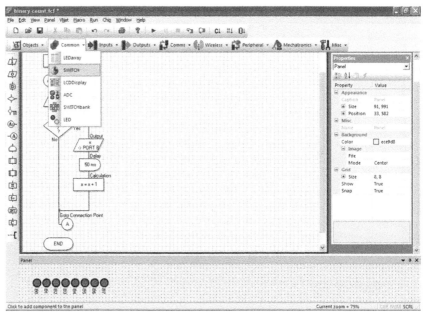

Figure 5.10

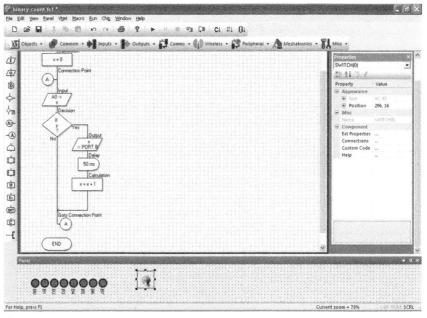

Figure 5.11

With the Toggle switch selected you can look at the properties of the switch in the Properties window. Look at Ext Properties and Connections; you don't need to change anything – leave them at their default values.

Running the Simulation

When you run the simulator, you can toggle the switch position by placing the mouse cursor on the Toggle switch and hitting a mouse button. The Toggle switch starts and stops the binary counting.

How the Program Works

We put the flowchart together without looking at the functions. The overview is that we are reading an input on one of the microcontroller's pins and, depending upon the electrical value (or binary value presented on the pin), the program will either count in binary displaying the numbers on the LEDs, or stop counting.

The switch, as shown in Figure 5.1, can deliver either a binary "1" or "0" to the input pin. The Input icon allows the program to read the I/O line (or pin) on Port A. Since we are only reading a single pin, we set the input properties of this icon to "Single bit." The value read at Port A is held in the variable "Y."

Our next icon, the Decision icon, reads the value "Y" and depending upon the value, branches the execution of the program to one of two branches. If the value of "Y" = "Yes" the program goes to the outside branch and continues to count. The values of "Yes" and "True" are accepted to mean binary "1." Consequently the values "No" and "False" are accepted to mean binary "0."

The Toggle switch component allows us to change the binary value present on the I/O line. To understand how the switch plays into the program, we need to look at the Toggle switch's properties. Select the switch and in the Properties window on the right select Ext Properties. The Edit Component Properties window opens as shown in Figure 5.12.

This gives basic information on our switch and its behavior. Close this window and open the Connections Properties window as shown in Figure 5.13.

In this property window we see the switch is connected to Port A Bit 0. We could change these connections but they are fine as they are. Close this window. All that's missing is how the

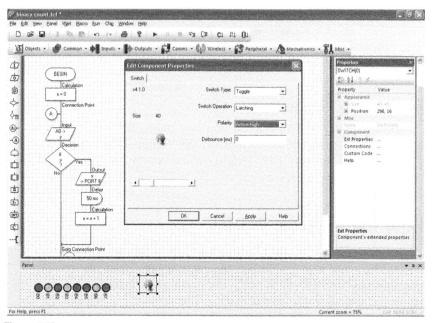

Figure 5.12

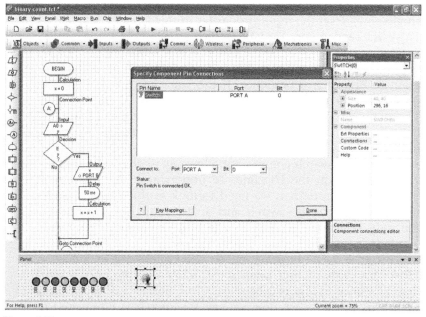

Figure 5.13

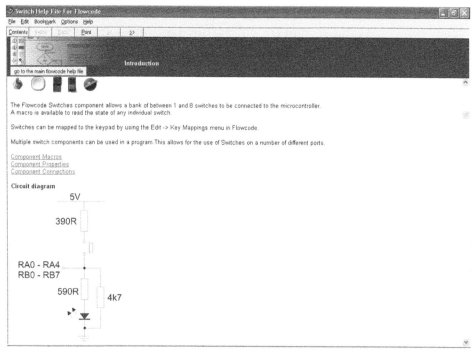

Figure 5.14

switch looks electrically inside the Flowcode developing environment. To see this, select "Help" in the Properties window. On the help page (see Figure 5.14) we have an electrical diagram of how the Toggle switch is represented electrically in the program.

While this switch circuit isn't identical to the switch circuit we are using, both switch configurations can deliver binary "1" and "0," depending upon the switch position, to the microcontroller I/O pin. If you haven't already, compile the program to a Hex file and load it into your microcontroller or ECIO28 and take it for a spin in the real world.

Alternatives

There is more than one way to program a function; in fact there are dozens. The way we just programmed that switch may not be the most efficient way, but I thought it explained the program's function and operation better. Let's look at an alternative.

Component Macros

In this section we will use Component macros. If you are not familiar with computers, your first question might be, "What's a macro?" A macro is a sequence of programmed instructions, a modular subprogram if you will, to be called upon from within a program when needed. The reason for macros is to make common programming tasks, like reading a switch that contains a number of program instructions, into an easy-to-use macro. So rather than writing the sequence of programming instructions to read a switch every time you wanted to read a switch, you would call a component macro instead. This makes programming components less tedious and less error-prone.

Flowchart Macros

We can create additional flowcharts within our workspace and call these newly created flowchart programs from other flowchart programs. Flowcode titles these additional flowcharts macros, and calling the flowchart program is a "Macro call." Readers who may have some programming experience will call these types of procedures "Subroutines." We will use Macro calls in later chapters. For now we will remain focused on Component Macros.

We can use a Component Macro to read our switch component instead of the Input icon. Before we change this program, use the File > Save As function to save this program with a different name or add a V2 to the end of the current name. This will preserve our current program. Once you have saved the file highlight the Input icon and hit the Delete key (see Figure 5.15).

Next select the Component Macro icon (see Figure 5.16) and drag it above the Decision icon, which is where the Input icon was before we deleted it in the flowchart.

Double click on the Component Macro icon to bring up its Properties window (see Figure 5.17). Inside the Properties window under Components, the two components in use in this flowchart are listed.

Select the SWITCH(0) component. Once the SWITCH(0) is selected the macros available for the SWITCH(0) are shown in the Macro's text box (see Figure 5.18).

In the Macro's text box, select "ReadState" then use the pulldown on the variable text box to select the "Y" variable (see Figure 5.19).

Then hit "OK." This writes the information in the "Call Component Macro" icon on the Flowchart (see Figure 5.20).

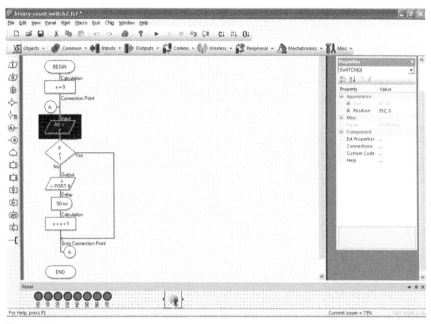

Figure 5.15

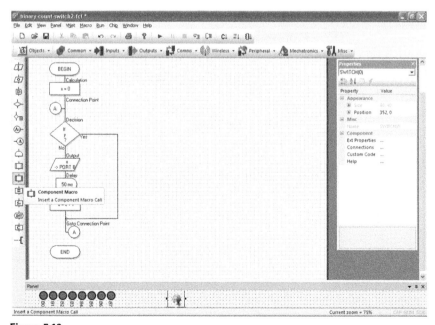

Figure 5.16

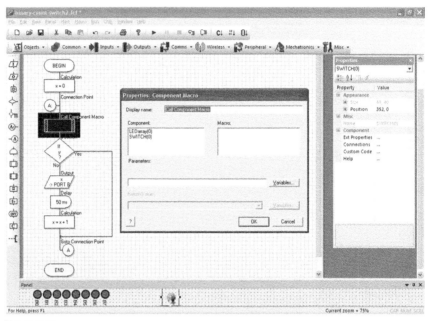

Figure 5.17

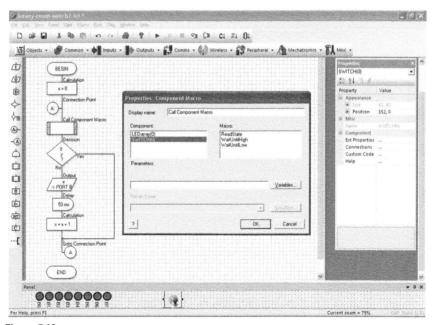

Figure 5.18

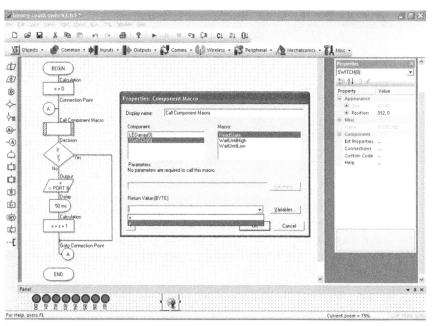

Figure 5.19

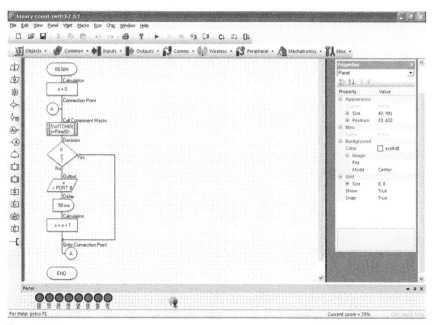

Figure 5.20

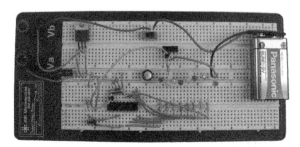

Figure 5.21

At this point you can run the simulation and use the Switch to control the counting. Once the simulation works you can have the Flowcode compiler generate a Hex file to program your microcontroller. Figure 5.21 shows the circuit on the solderless breadboard.

6

LANGUAGE REFERENCE GUIDE

In this chapter we will provide an overview of the Flowcode icons available for use. The icons are listed in the order that they appear in the toolbar.

Input Icon

Toolbar

Icon (Figure 6.1).

Flowchart icon (Figure 6.2).

Input icons check the specified port and/or bits for their value and then place the resulting value into the specified variable (Figure 6.3).

- Display Name is the name labeled for the icon when it appears on the flowchart.
- Variable text box allows you to select the name of the variable that you wish to input the port state into.
- The Variables button brings up the Variables dialog window allowing you to select an existing variable or to create a new one.

Figure 6.1

Figure 6.2

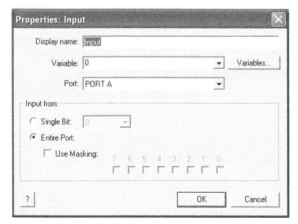

Figure 6.3

PIC Projects for Non-Programmers. DOI: 10.1016/B978-1-85617-603-3.00006-3

- The Port text box allows you to select from a list of the ports available on the target microcontroller.

Input: Single Bit

Use this option to read the state of a single bit of the port. When the state of a single bit is true the value of that bit is passed as its value.

Input: Entire Port

Use this option to read the value of the entire port into the variable.

Use Masking

When masking is used, it is possible to read only certain bits into a variable. This is useful when some bits of the port are configured as outputs. When a mask is used only the values of the selected bits are read. See Flowcode Help for more detail.

Output Icon

Figure 6.4

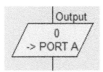

Figure 6.5

Toolbar

Icon (Figure 6.4).

Flowchart icon (Figure 6.5).

Output icon is used to send the value or variable to the specified port and/or bits. The output is received at the port in binary format. For instance, sending the number 3 to an 8-bit output port will appear in binary as 00000011 at the port.

Properties window (Figure 6.6).

Display Name is the name labeled for the icon when it appears on the flowchart. Variable or value text box allows you to select the name of the variable or a numeric value that you wish to output to the port. Numeric values can be in decimal (unmarked) or in Hex format (preceded by $0 \times$) e.g., 255 or $0 \times FF$.

Variables Button

This button brings up the Variables dialog window allowing you to select an existing variable or to create a new one.

Port

Select from a list of the ports available on the target microcontroller.

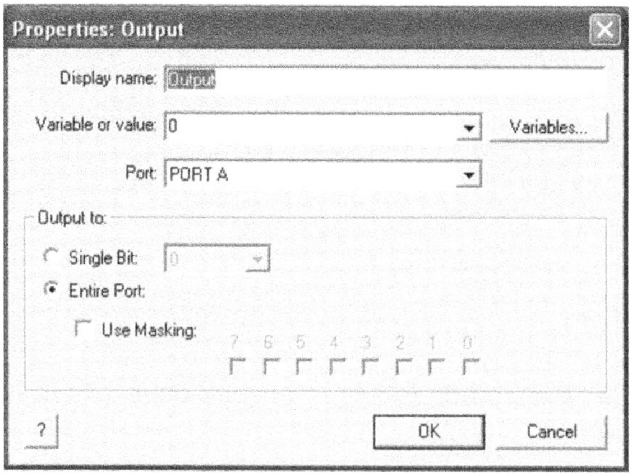

Figure 6.6

Single Bit

Use this option to write to a single bit of the port. If a true (non-zero) value is sent to the bit the bit is set (turned on), otherwise the bit is cleared (turned off).

Entire Port

Use this option to write the variable or value to the entire port.

Use Masking

When masking is used, it is possible to write to only certain bits of a port. This is useful when some bits of the port are configured as inputs and you wish to leave these bits unchanged. With masking only the selected bits receive their value; all non-selected bits are not affected. See the Using Masks page for more detail.

Delay Icon

Toolbar

Icon (Figure 6.7).
Flowchart icon (Figure 6.8).
Delay icons are used to slow down program execution and for program timing. They are particularly useful in slowing program execution speed down to allow for human interaction (Figure 6.9).

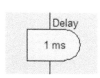

Figure 6.7

Figure 6.8

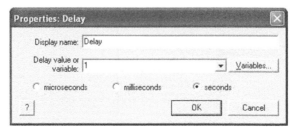

Figure 6.9

Display Name is the name labeled for the icon when it appears on the flowchart.

Delay Value or Variable

This is the length of the delay that you wish to create.

Variables Button

This button brings up the Variables dialog window allowing you to select an existing variable or to create a new one.

Microseconds/Milliseconds/Seconds Options

Delays can be specified in units of microseconds, milliseconds or seconds. When simulating delays in seconds, a dialog box will appear showing how much of the delay has elapsed. A cancel button on the dialog allows the execution of the flowchart to continue before the entire delay has passed. To allow Flowcode to correctly program your chip with the correct delay setting you will need to select a clock speed for your chip. The dialog box for this can be seen by selecting Edit ... Project Options.

Switch Icons

Toolbar

Icon (Figure 6.10).
Flowchart icon (Figure 6.11).
Switch icons are similar to decision icons, but instead of two possible outcome branches (Yes, No) there are up to eleven. They are particularly useful for tailoring your program to react in different ways based on an input variable (Figure 6.12).

Figure 6.10

Figure 6.11

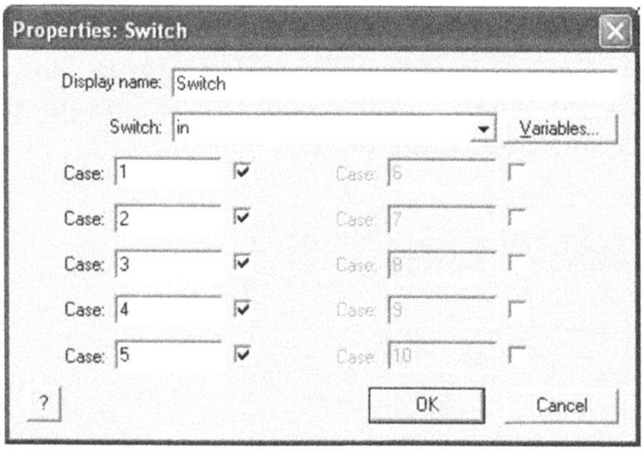

Figure 6.12

Display Name is the name labeled for the icon when it appears on the flowchart.

Switch

This is the input parameter variable that is used to determine the branch of the switch that will be executed at run time.

Variables Button

This button brings up the Variables dialog window allowing you to select an existing variable or to create a new one.

Case Options

Up to ten branches can be defined for the switch icon. The switch branches are enabled by ticking the boxes next to the values. The value in the case box is the value that is used to trigger that particular branch of the switch icon. If none of the values matches the input parameter variable then the default branch of the switch case is used.

Connection Point 1

Figure 6.13

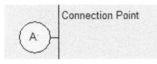

Figure 6.14

Toolbar

Icon (Figure 6.13).
Flowchart icon (Figure 6.14).
Connection icons are used to link one part of a flowchart to another. When the flowchart reaches the jump point it jumps to the matching connection point and then continues execution. Connection icons are used in pairs; part one is the connection point – the point in the flowchart to jump to. Part two is the jump point – the point in the flowchart to jump from. Both parts share a connection letter – in this case "A." Several jump points can reference one connection point (Figure 6.15).

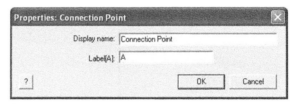

Figure 6.15

Display Name

This is the name of the icon that appears on the flowchart.

Label

This is the text that will be displayed on the connection point allowing for meaningful labels to be created.

Figure 6.16

Figure 6.17

Part Two: The Jump Point

Toolbar

Icon (Figure 6.16).
Flowchart icon (Figure 6.17).

Loop Icon

Figure 6.18

Toolbar

Icon (Figure 6.18).
Flowchart icon (Figure 6.19).

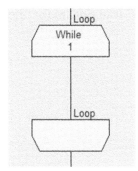

Figure 6.19

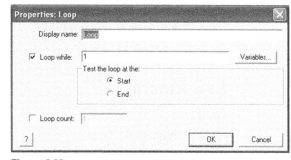

Figure 6.20

Loop icons are used to repeat a task until specified conditions are fulfilled, or to perform the loop a set number of times. Note that you will need to add calculation icons that modify the variables used for the condition in order for the condition to be fulfilled (Figure 6.20).

Display Name is the name labeled for the icon when it appears on the flowchart.

Loop While

This loops the program until the specified condition becomes true. Enter the conditions which will fulfill the loop. (Setting the test condition to something that is always true will make the loop repeat forever, e.g., While 1.)

Variables Button

This button brings up the Variables dialog window allowing you to select an existing variable or to create a new one.

Test the Loop at the . . .

Select whether you wish the loop to be tested at the start or at the end of the loop. This can be set to check the condition at the start of the loop or at the end of the loop.

Loop Count

This sets the loop to loop through a set number of times. The count must be a whole number between 1 and 255.

Infinite Loops

Sometimes a task needs to be repeated continually. One useful way of achieving this is to have an infinite loop. Setting the test condition to something that is always true will make the loop repeat forever, e.g., While 1.

Call Macro

Toolbar

Figure 6.21

Icon (Figure 6.21).

Flowchart icon (Figure 6.22).

Macros have now been split into Macros and Component macros. Component macros are predefined macros supplied with the components. For example, LCD macros are used to display numbers and characters on the LCD display. Component macros are only available for use with that particular component. Component macros have hatched outer bands on the icon.

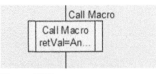

Figure 6.22

Macros (explained here) are those designed and created by the user. Macros can be exported and imported allowing users to build up macro libraries for common or important tasks. On macros the outer bands are clear and not hatched (Figure 6.23).

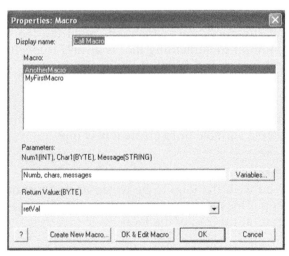

Figure 6.23

Macros

Macros are sections of code that can be used and reused in projects. They allow complex tasks to be handled by code blocks and they can be imported and exported. The macro properties dialog allows users to select and create macros.

Adding a Macro

Select the macro to use from the list, or click on Create New Macro... to begin creating a new macro to add to the list. Add in any parameters required, and select a return value if required. Clicking on the OK button completes the process and closes the dialog box. Clicking on the OK and Edit button completes the process, closes the dialog and opens the selected macro for editing.

Parameters

If the device macro requires any parameters then these can be entered in this field. They can be numeric values or existing variables. Each variable or value must be separated by a comma. The parameter details will list the parameter type. Parameters must be of the required type to be accepted. Note that a full set of parameters must be provided.

Variables Button

This button brings up the Variables dialog window allowing you to select an existing variable or to create a new one.

Return Value

If the device macro returns a value then you can assign that value to an existing variable for use later in the flowchart. If the function returns a value but you do not wish to retrieve it then leave this field empty. The variable type of the return value will be listed. A variable of the specified type must be used to accept the return value.

Create New Macro Button

Select this button to create a new macro within the flowchart.

OK and Edit Macro Button

Clicking this button will open up the chosen macro to allow it to be viewed or edited.

Creating New Macros

Clicking the Create New Macro... button brings up the new macro dialog. See "Creating new macros", page 134 for more information.

Component Macros

Toolbar

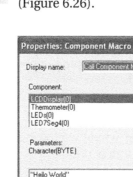

Figure 6.24

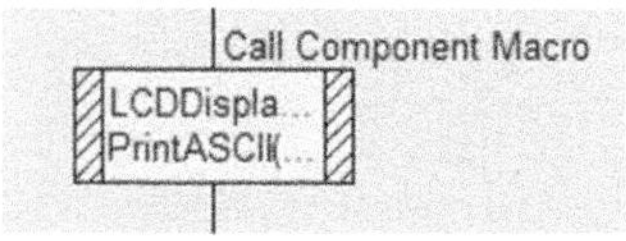

Icon (Figure 6.24).

Flowchart icon (Figure 6.25).

Macros have now been split into Macros and Component macros. Component macros are pre-defined macros supplied with the components. For example, LCD macros are used to display numbers and characters on the LCD display. Component macros are only available for use with that particular component. Component macros have hatched outer bands on the icon (Figure 6.26).

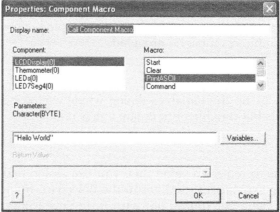

Figure 6.25

Figure 6.26

Macros are those designed and created by the user. Macros can be exported and imported allowing users to build up macro libraries of common or important tasks. On macros the outer bands are clear and not hatched.

Adding a Component Macro

Select the component and macro to use. Select the component from the list of attached components on the left, and select the macro to use from the list on the right. Add in any parameters required, and select a return value if required.

Parameters

If the device macro requires any parameters then these can be entered in this field. They can be numeric values or existing variables. Each variable or value must be separated by a comma.

The parameter details will list the parameter type. Parameters must be of the required type to be accepted. Note that a full set of parameters must be provided.

Variables Button

This button brings up the Variables dialog window allowing you to select an existing variable or to create a new one.

Return Value

If the device macro returns a value then you can assign that value to an existing variable for use later in the flowchart. If the function returns a value but you do not wish to retrieve it then leave this field empty. The variable type of the return value will be listed. A variable of the specified type must be used to accept the return value.

Calculation Icon

Toolbar

Icon (Figure 6.27).

Flowchart icon (Figure 6.28).

Calculation icons allow the modification of variables. They can be used to check inputs and to create outputs (Figure 6.29).

Figure 6.27

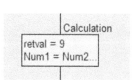

Figure 6.28

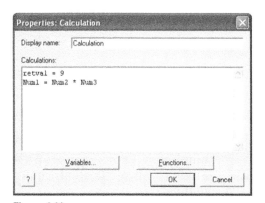

Figure 6.29

Display Name is the name labeled for the icon when it appears on the flowchart.

Calculations

One or more lines of calculations can be entered into this box. All calculations must consist of the name of an existing variable, an equals sign and an expression made up from numbers, variables and the following operators:

( )	: Parentheses.
= <>	: Equal to, Not equal to.
± * /MOD	: Addition, Subtraction, Multiplication, Division and Modulus.
< <=>	: Less than, Less than or equal to, Greater
>=	than, Greater than or equal to.
>> <<	: Shift right, Shift left.
NOT AND OR XOR	: NOT(inversion), AND, OR, Exclusive OR.

Numeric values can be in decimal (unmarked), in Hex format (preceded by 0×), or in binary format (preceded by 0b), e.g., 85 or 0 × 55 or 0b01010101. Assuming that the variables have been previously defined, all the following are valid calculation lines.

DELAY = DELAY + 1
DELAY = (MYVAR + 3) * 3
NEXTBIT = LASTBIT >> 2 & MASK
AANDB = PORT_A AND PORT_B
INVX = NOT X

Browse for Variable Button

This button brings up the Variables dialog window allowing you to select an existing variable or to create a new one.

Functions (Flowcode PIC Only)

Flowcode includes an additional set of mathematical functions:

float = fadd(float, float)	Add two floating point numbers together
float = fsub(float, float)	Subtract two floating point numbers
float = fmul(float, float)	Multiply two floating point numbers
float = fdiv(float, float)	Divide two floating point numbers
float = fmod(float, float)	MOD function for floating point numbers
byte = isinf(float)	Checks to see if the floating point number is infinite
byte = isnan(float)	Checks to see if the floating point is not a number

byte = float_eq(float, float)	Compares two floating point numbers to see if they are equal
byte = float_ge(float, float)	Compares two floating point numbers to see if they are greater than or equal
byte = float_gt(float, float)	Compares two floating point numbers to see if they are greater than
byte = float_le(float, float)	Compares two floating point numbers to see if they are less than or equal
byte = float_lt(float, float)	Compares two floating point numbers to see if they are less than
int = random()	Generates a random number $-32768 <=> 32767$

String Icon

Toolbar

Icon (Figure 6.30).

Figure 6.30

Flowchart icon (Figure 6.31).

The Properties box of the String manipulation icon allows users to manipulate, create and edit strings in a similar manner to the way the calculation icon allows users to manipulate numeric variables (Figure 6.32).

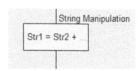

Enter the function code and string variable names into the text box to manipulate the strings. The Variables and Function buttons allow the user to add those elements into the edit box window.

Figure 6.31

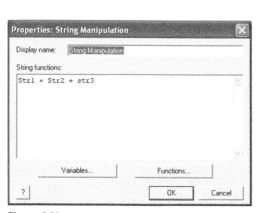

Figure 6.32

Strings

Strings are arrays of BYTE ASCII character values. String arrays are composed of the array name and an array element variable. MyString[24], for instance, is an array called MyString that is 24 characters in size (Figure 6.33).

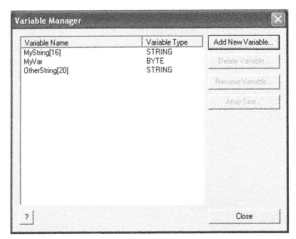

Figure 6.33

Creating a String

See Figure 6.34.

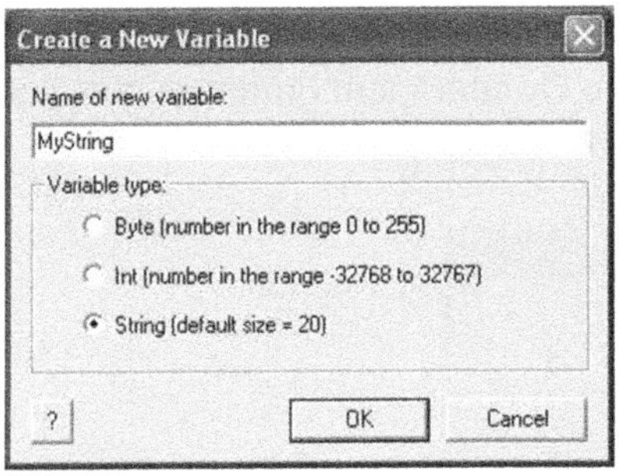

Figure 6.34

Strings are created in the Variables screen along with other variables. To create a string, enter the string array name and tick the string option. To set the array size, enter the desired size inside square array brackets. If no size is specified the string will be created with the default size of 20 characters (Figure 6.35).

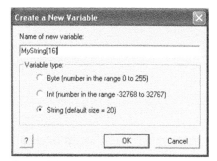

Figure 6.35

Resizing the String

To resize the string open the variables window, select the string to resize and select Edit. Enter the new size for the array (Figure 6.36).

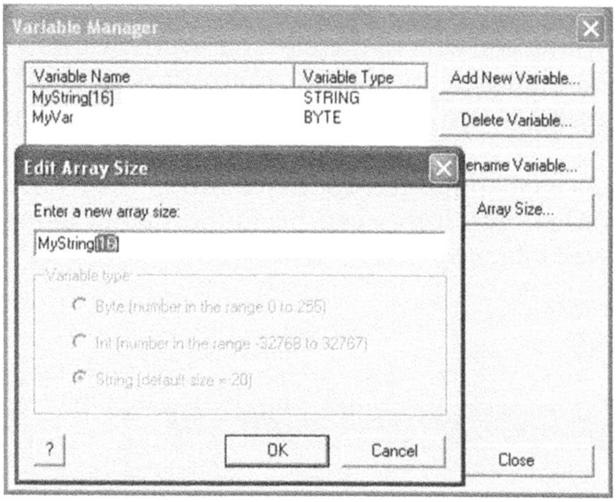

Figure 6.36

Note that you may need to alter your program code to accommodate the size change.

String Manipulation Functions

The string functions are a set of string manipulation functions that can be used to edit, change and examine the strings. Clicking on a function adds the base code to the edit box window. The user can then edit this base code with the variables required.

Figure 6.37

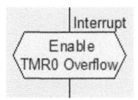

Figure 6.38

Interrupt Icon

Toolbar

Icon (Figure 6.37).
Flowchart icon (Figure 6.38).

Introduction to Interrupts

Interrupts are called to react to an event such as an external stimulus or a timing event. When an interrupt is activated it executes a macro. The macro to be called is specified by the user when creating the interrupt. The number and types of interrupts available are dependent on the device used. Some devices have quite a number of interrupts, whereas other devices may only have a few.

The exact details, properties and operation of an interrupt vary from interrupt to interrupt. Users will need to refer to the interrupt dialogs for exact details of any particular interrupt. However, there are four main types that Flowcode uses:

- TMR<X> – overflow – reacts to the clocked overflow event.
- INT – reacts to an input on an external interrupt pin.
- Port change – reacts to any change in a specified collection of inputs.
- Custom – defines a custom interrupt procedure.

Details of the various interrupts can be found below, along with examples of available properties (Figure 6.39).

Creating an Interrupt

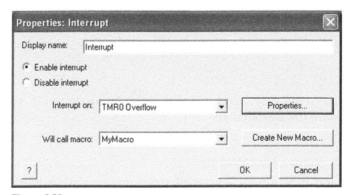

Figure 6.39

Add an interrupt icon to the program. Select Enable interrupt or Disable interrupt to turn the interrupt on or off respectively. Select the desired interrupt from the drop-down list of interrupts available from that particular device. Set the properties for that event. Set the macro to call, or create a new macro.

Interrupt Properties

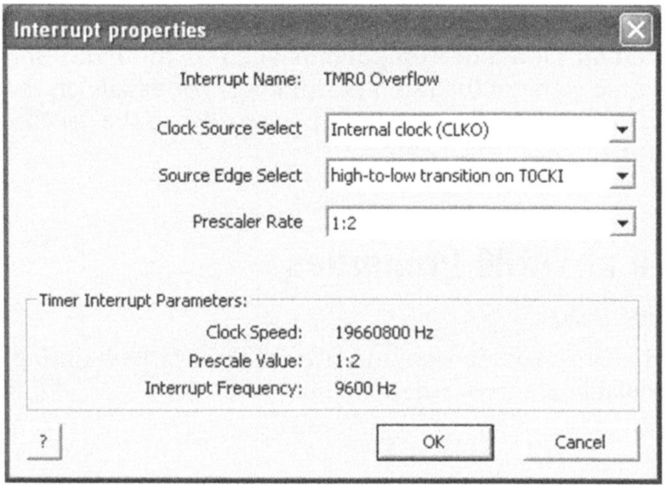

Figure 6.40

TMR<X> Timer overflow interrupt (e.g., TMR0 overflow). Interrupts when the timer overflows. A count is made internally based on clock inputs and the prescaler property (see below for details). When the count reaches a certain point it overflows back to zero triggering an overflow event. This event can then be used to call a macro. Timer overflow interrupts are called repeatedly at the interrupt time interval, making them very useful for time based events or for frequently needed features such as display updates.

Note: Check that the clock speed setting is correct (via the Chip → Clock Speed menu items) as clock speed affects timer interrupt frequency settings.

The exact properties available for a timer interrupt are device- and timer-dependent. As such, the timer overflow dialogs may vary from device to device and from timer to timer. The example below is intended to give an idea of the kinds of properties available for a timer overflow interrupt. Specific details can be found in the Interrupt dialogs, and from device datasheets. TMR1 on the 16F877A, for instance, has a smaller set of prescaler options than the TMR0 interrupt and also has no source edge settings available, unlike the TMR0 interrupt.

The TMR2 interrupt has an additional setting called the rollover value. This changes the rollover value in the timer register that is responsible for triggering the interrupt, which allows for much better control over interrupt frequency. This advanced timeout functionality can be replicated with the other timers by writing to the timer count registers directly using C code.

Note: Certain Flowcode components such as the PWM and Servo consume some of the timer peripherals. For details on the hardware consumed by a component please refer to the specific component help file.

Example of TMR0 Properties

Clock Source Select

Select the clock source signal to use for the overflow timing. Options available are device-dependent.

Examples
- Internal clock (CLK0)
- Transition on T0CKI.

Source Edge Select

Select the transition edge used for clock signal timing.
- High-to-low transition on T0CKI
- Low-to-high transition on T0CKI.

Prescaler Rate

Select a prescaler rate for the clock signal. A prescaler divides down the clock signals used for the timer, giving reduced overflow rates. The rate can be set to a number of possible values. The exact values are chip-dependent; e.g., for the PIC16F877A values area a number of values are available ranging from 1:1 to 1:256. The prescaler value is used in conjunction with the clock speed to set an overflow frequency, e.g.:
- Clock speed: 19660800
- Prescale value: 1:256
- Interrupt frequency 75 Hz
- Clock speed: 19660800

- Prescale value: 1:64
- Interrupt frequency 300 Hz.

INT

This is triggered when an external stimulus is used on an external interrupt (INT) pin (e.g., a button press). The interrupt can be set to trigger on the:
- Falling edge of INT
- Rising edge of INT.

INT interrupts can be useful in setting interrupts to occur on signal-on or signal-off events, such as an emergency shut off switch.

Port Change

This interrupts when an input signal occurs on any of a selected group of port inputs. The port change interrupt may have properties allowing the interrupt I/O pins to be enabled or disabled as interrupt pins.

Custom

The Custom option allows the creation of custom interrupt code. Many of the chips available in Flowcode support hardware driven interrupts that are not provided in the standard set of Flowcode interrupts. To allow users to add their own interrupts to suit their application Flowcode has the option to add your own custom interrupt.

Interrupt Properties Page

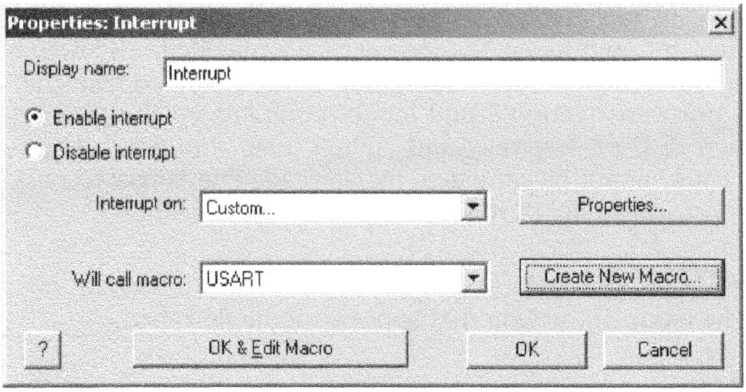

Figure 6.41

Custom interrupt properties page.

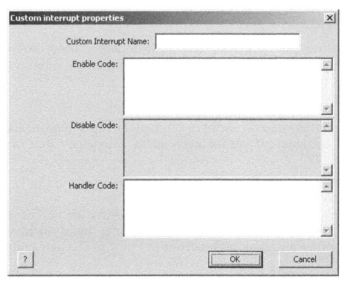

Figure 6.42

Code Icon

Toolbar

Figure 6.43

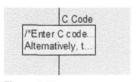

Figure 6.44

Icon (Figure 6.43).

Flowchart icon (Figure 6.44).

Flowcode is designed to allow those new to chip programming to design programs with little knowledge of other programming languages. However, it is possible to embed code written in other languages into Flowcode programs. Programs written in C and Assembly code can be embedded in Flowcode using the Code icon. This means that it is possible to take complex programs (you will find lots of programs on the Internet) written in C or Assembly and embed them into your designs. The code cannot be simulated by Flowcode, but is passed on to the microcontroller during compilation (Figure 6.45).

Display Name

The name of the icon that appears on the flowchart.

C Code

Enter any C code you wish to include in the flowchart. The C code is not checked by Flowcode but is passed straight to the

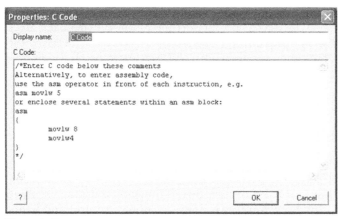

Figure 6.45

C compiler when the flowchart is compiled. It is important to verify that the C code entered is correct, as syntax errors will cause the compilation of the whole flowchart to fail.

To access Flowcode variables, macro functions and connection points, it is necessary to color the variable in your C code with the prefixes FCV_, FCM_ and FCC_MacroName_ respectively. For example, to use a Flowcode variable called DELAY in your C code, you must refer to it using FCV_DELAY. Note that all Flowcode defined variables are upper case. To call a Flowcode macro called TEST in your C code, you must call FCM_TEST();. Note that all Flowcode macro names are upper case.

To jump to a connection point called A, defined in a Flowcode macro called TEST, your C code must be goto FCC_TEST_A;. Connection points defined in the main flowchart of a Flowcode file are prefixed FCC_Main_. To enter a tab character in the C code field, use Ctrl + Tab.

Assembly Code

Assembly code can be added into the code field in a C assembly code wrapper. For a single line of code use the asm operator in front of each instruction, e.g., asm movlw 5, or enclose several statements within an asm block:

```
asm
{
; Enter your code here
}
```

To access Flowcode variables, macro functions and connection points, it is necessary to further color the prefixes used above with a preceding underscore, _FCV_, _FCM_ and _FCC_MacroName_ respectively. The examples used above now become _FCV_DELAY, _FCM_TEST() and _FCC_TEST_A.

Comment Icon

You can add comments to your code with the Comments icon. Drag the icon to where you want it and add your comments in the icon's Properties box. Note that the icon is off to one side of the flowchart, as it does not need running and is not a part of the flowchart as such. Comments allow users to leave notes in their code explaining functions, variable settings, etc., or for any other documentation task that they feel is needed. Leaving comments is a good idea as it helps others understand your code.

7

LIQUID CRYSTAL DISPLAY (LCD)

One big disadvantage of microcontrollers is that they do not have an output display. So far we have only been able to see a binary output of our microcontroller using LEDs. How much better would it be to connect an alphanumeric display? This would allow our microcontroller to communicate using text messages and numbers. An alphanumeric display will make our microcontroller projects far more user friendly. That's the focus of this chapter.

The LCD module we will use needs to be compatible with the Hitachi 44780 controller. Most LCDs manufactured today are compatible with this controller, so this should not be an issue. These LCDs have a 14- or 16-pin connection. The connection may be a single or double row header at one edge of the LCDs' printed circuit board (pcb) (see Figure 7.1).

The LCD module we are using is sixteen characters by two lines (16 × 2). To use the LCD you first need to build a circuit. To build the circuit you need to obtain an LCD module, with its data sheet. The LCD will use I/O pins that are common to LCDs in general. The datasheet will tell you the pin numbers of these I/O pins that you will need to build the circuit. Figure 7.2 is a diagram of what you may find inside an LCD data sheet identifying the I/O pins.

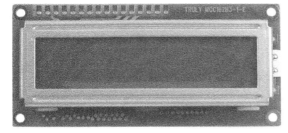

Figure 7.1

PIC Projects for Non-Programmers. DOI: 10.1016/B978-1-85617-603-3.00007-5

■ EXTERNAL DIMENSIONS

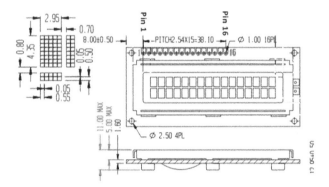

■ BLOCK DIAGRAM

1	2	3	4	5	6	7	8	9	10	11	12	13	14	15	16
VSS	VDD	VO	RS	R/W	E	DB0	DB1	DB2	DB3	DB4	DB5	DB6	DB7	NC	NC

Figure 7.2

LCD Data Pins

The LCD typically has an 8-bit data bus labeled D0 to D7. We only need to use four of these data lines. If you look at the LCD module Help screens in Flowcode, they show a schematic that uses the lower four data pins, D0 to D3. I have not been too successful using the lower data pins. I have better success using the upper four data pins D4 to D7 and the schematic that follows reflects this choice. The other LCD pins that you will need are:

- Vss ground
- Vdd positive power (+5 volts)
- Vo contrast control
- R/S register select
- R/W read write
- E enable.

Once you have these pin numbers identified from your LCD data sheet we are ready to begin.

Make Your LCD Module Breadboard Friendly

A 16-pin header soldered to the data pin holes on the LCD printed circuit board allows you to plug the board into the solderless breadboard (see Figures 7.3 and 7.4).

We can then connect wires to the LCD data pins. Figure 7.5 shows the LCD schematic.

Figure 7.3

Figure 7.4

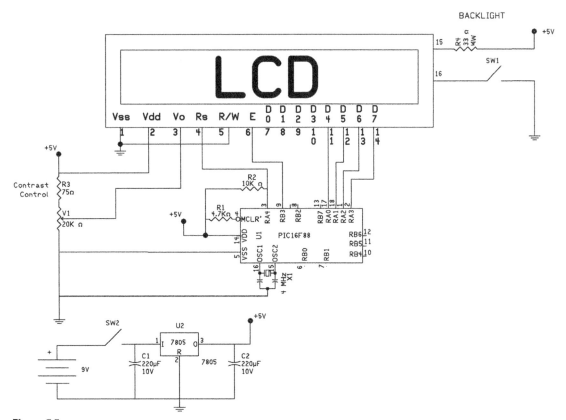

Figure 7.5

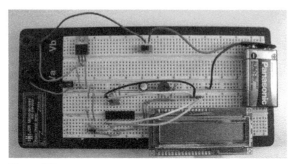

Figure 7.6

Build this schematic onto the solderless breadboard. When you are finished it will look something like Figure 7.6.

Let's Start Programming

Start a new programming screen and choose the 16F88 microcontroller (see Figure 7.7).

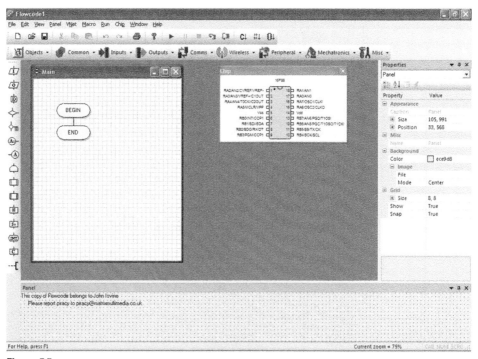

Figure 7.7

Save the program as "LCD_Test." Before we start looking at programming techniques for the LCD, we want to be sure our circuit works. So to test our circuit we will keep the program to the bare essentials to check the circuit and then continue with more advanced programming.

Step 1:

Add the LCD module. Go to the Common menu item and select LCD Module (see Figure 7.8).

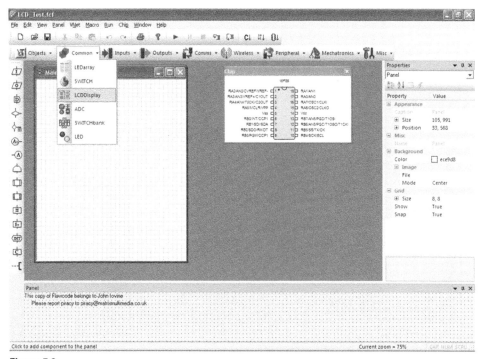

Figure 7.8

We have to assign which microcontroller pins we are connecting to which LCD pins. The default values will not work for us. Select the LCD in the bottom panel. Then go to the Properties window on the left. Select "Connection" under the Component Submenu item. Once "Connections" is highlighted, click on the three dots "..." to the right to open up the "Specify Component Pin Connections" window. Change the port pins to:

Pin Name	Port	Bit
Data 1	Port A	0
Data 2	Port A	1
Data 3	Port A	2
Data 4	Port A	3
RS	Port A	4
Enable	Port B	3

This is shown in Figure 7.9.

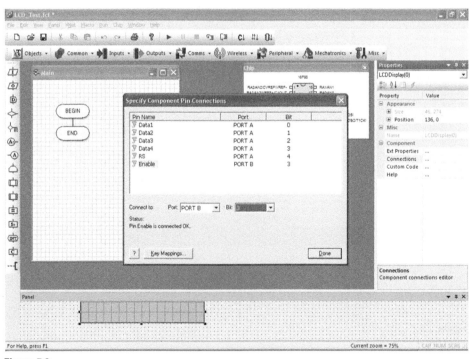

Figure 7.9

Step 2:

Add C Code. We need to add one line of code to turn off the analog lines on the 16F88 microcontroller. This allows those pins to be used as digital I/O. Add the "C" code icon to the Flowchart. Open up its Properties window. Erase all the comments that are listed inside and then add the following line:

ansel = 0;

Don't forget the ";" at the end of the line or the command will not work (see Figure 7.10).

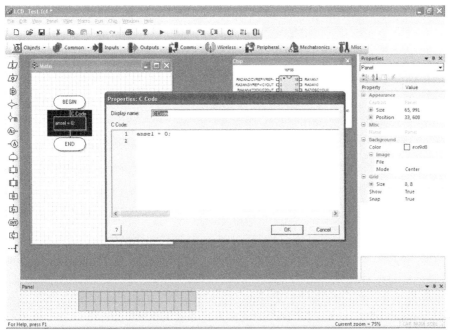

Figure 7.10

Step 3:

Add components macro. Select the Component Macro icon on the left and drag it onto the Flowchart (see Figure 7.11).

Double click on the inserted Component Macro icon to open its Properties window. The window will display the available components in the Flowchart. In this case you only have one, the LCD (see Figure 7.12) identified as LCDDisplay(0).

When you select and highlight the LCDDisplay(0), the macro available for use with the LCD is displayed (see Figure 7.13).

Select the "Start" macro, which is needed to initialize the LCD for use, and click OK.

We will now add a few more "Component Macro" icons to the chart and activate different aspects of the LCD control. For the next component choose the "PrintACSII" macro and write "Hello World..." in the variable space and click OK (see Figure 7.14).

The next call will position the LCD cursor on the LCDs second line in the first position. Since the LCD starts counting at

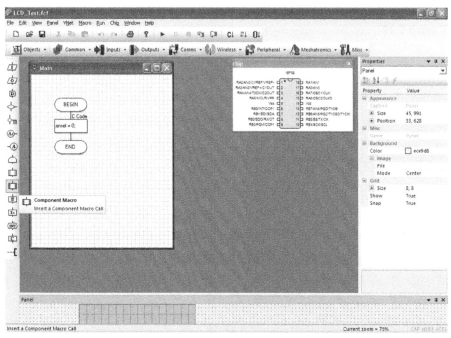

Figure 7.11

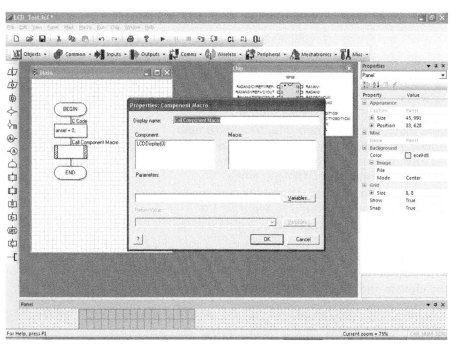

Figure 7.12

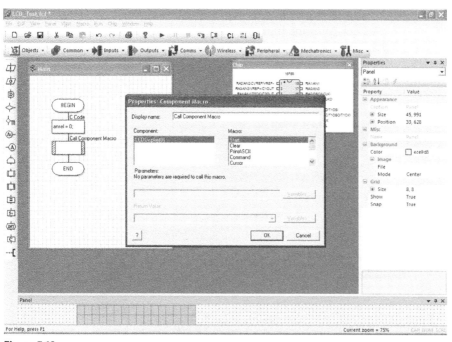

Figure 7.13

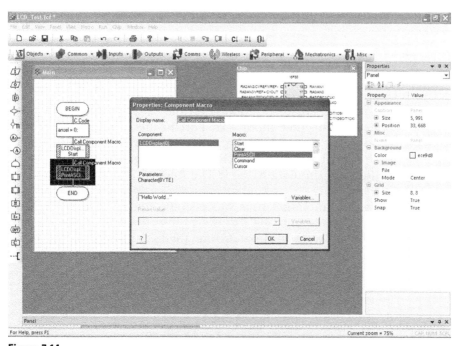

Figure 7.14

")", the first position is "0" and the second line is "1." For the next component macro choose "Cursor" macro and write 0,1 in the variables space (see Figure 7.15).

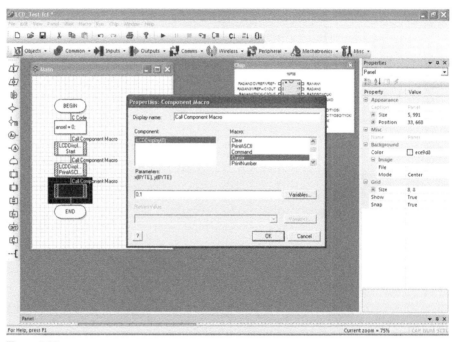

Figure 7.15

With the cursor positioned on the second line in first position we can enter more text. For the next component macro choose the "PrintACSII" macro again and write "Testing 1..2..3," in the variables space and click OK (see Figure 7.16).

Run the simulation; the LCD output should appear as shown in Figure 7.17.

If the simulation runs successfully, compile the program to Hex. Load the Hex file (firmware) into the EPIC software, and set the software to program the 16F88. You need to make the following changes to the EPIC configuration screen. Change:

a. Oscillator to XT

b. Watchdog Timer to Disabled

c. Power-up Timer to Enabled

d. Brown-out Reset to Disabled

e. Low Voltage Programming to Disabled

f. Flash Program Memory to Enabled (see Figure 7.18).

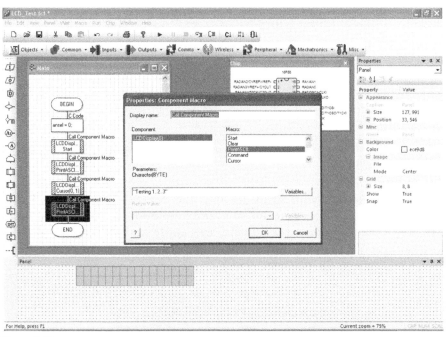

Figure 7.16

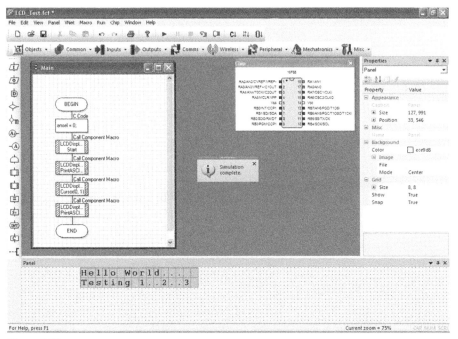

Figure 7.17

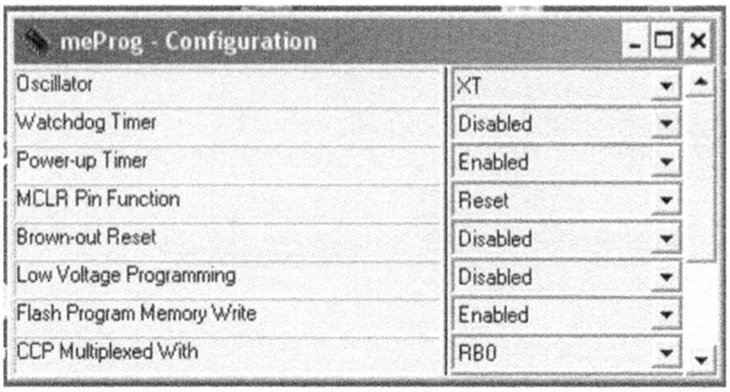

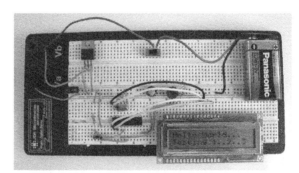

Figure 7.18

Upload your firmware into the 16F88 using the EPIC programmer. Next insert the programmed microcontroller into your solderless breadboard and apply power (see Figure 7.19). As a side note, the Matrix Multimedia program has similar settings on its interface as EPIC.

Figure 7.19

Troubleshooting the Circuit

If, when you first turn on your circuit you don't see anything printed on the LCD, don't panic. Your first step in troubleshooting is to adjust the potentiometer. The potentiometer controls the contrast on the LCD screen and if it is not adjusted properly it could easily hide the text. Adjust your potentiometer. If that doesn't work, recheck the wiring of your circuit.

Scrolling Text

Now that we can print messages on the LCD screen, let's manipulate the text by scrolling it back and forth. Figure 7.20 shows the Flowchart for scrolling the text.

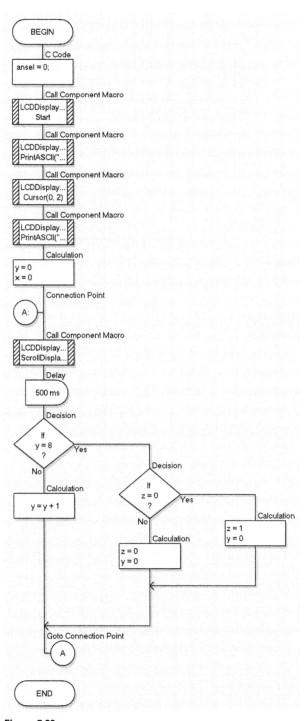

Figure 7.20

We used just about every icon and technique illustrated in Figure 7.20. However, in the Calculation icon about the A: Connection point, create the following byte variables; y, x and z. Then assign the following values to the variables by entering the following data: y = 0, x = 0 and z = 1. Underneath the A: Connection point, the Call Component Macro uses the Scroll Display Macro (see Figure 7.21).

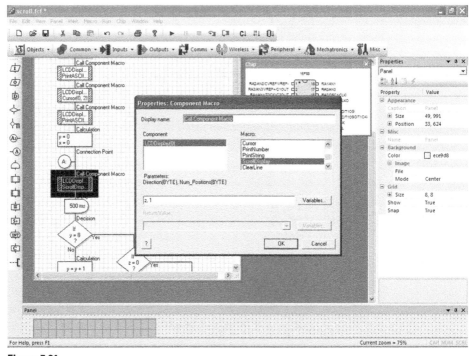

Figure 7.21

Under the macro parameters enter two variables: Direction (BYTE) and Num_Positions(BYTE). The Direction byte can have a value of 1 to scroll to the right or 0 to scroll to the left. The Num_Positions(BYTE) determines the spaces the text scrolls. Enter the variables z,1. You have used all the other icons previously in this book so I'm leaving it up to you to use the flowchart diagram to enter the balance of the program.

Positioning the Cursor

While we positioned the cursor to the second line first position in the previous programs, the cursor may be positioned anywhere on the LCD screen. Each LCD character's "on-screen" position x,y coordinates are illustrated in Figure 7.22.

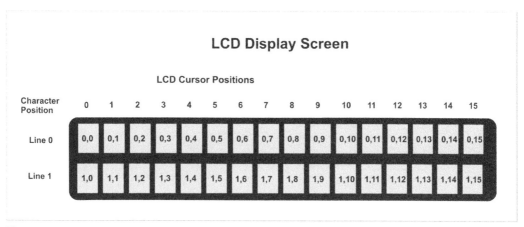

Figure 7.22

For example, to position the cursor on the second line seventh position, insert the Call Component into your flowchart, choose "Cursor" macro and write 1,7 in the variables space, as was illustrated in Figure 7.15.

Off-Screen Memory

Each line on the LCD holds 40 characters; only the first 16 characters are displayed on the LCD screen. To test this we will use our scroll commands to view text entered past the display character spaces. Figure 7.23 is the new Flowchart we are working with. It is similar to Flowchart 7.20 so we can use that as a base. Resave the program using a new name. Then there are a few changes we need to perform.

a. In both the LCD Display PrintASCII, add 16 empty spaces in front of the existing text.

So that "Hello World.." becomes "   Hello World.." and

"Testing 1..2..3.." becomes "   Testing 1..2..3"

b. In the Calculation icon about the A: Connection point, assign the following values to the variables by entering the following data: $y = 0$, $x = 0$ and $z = 0$.

c. In the Decision icon "If $y = 8$" is changed to "If $y = 15$"

Make these changes and when you run the simulation, the hidden text will scroll from the off-screen memory to the on-screen memory, then continue to scroll back and forth (see Figure 7.23).

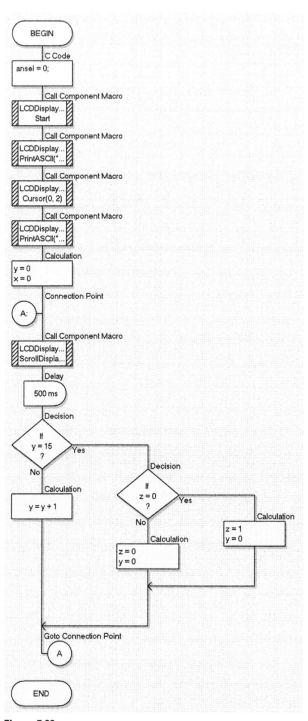

Figure 7.23

Printing Numbers

You will have occasions where you need to print out the numeric value of variables. The LCDDisplay(0) component has a Printnumber macro. We use this macro in the following Flowchart to print the value of x on the LCD screen (see Figure 7.24).

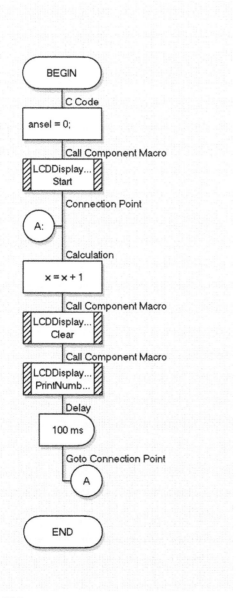

Figure 7.24

This flowchart requires you to define a byte variable x. The PrintNumber macro is shown in Figure 7.25.

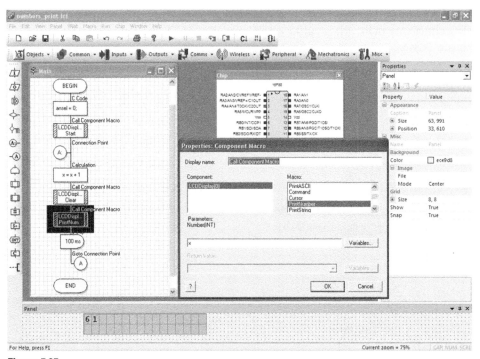

Figure 7.25

Printing Numbers at Specific Locations

The following flowchart (Figure 7.26) uses the Cursor macro to print the x variable approximately center screen on the first line of the LCD.

Creating Macro Messages

Embedding multiple LCD messages inside your main Flowchart can make the chart become long and tedious. One way to simplify LCD messages is to place the messages in smaller satellite Flowcharts called macros. These messages may then be called at anytime from anywhere within your main Flowchart. If you have programmed before, calling smaller flowcharts is similar to creating and calling subroutines.

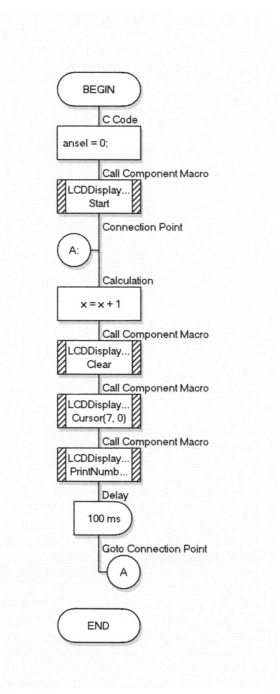

Figure 7.26

To Create a New Macro

Select Macro → New, as shown in Figure 7.27.

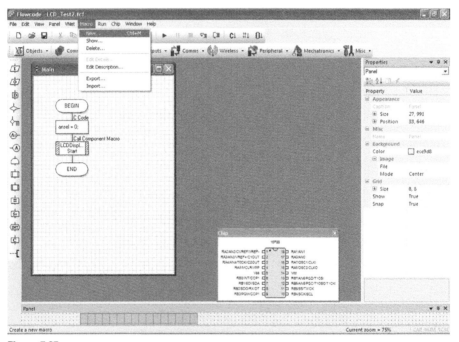

Figure 7.27

This opens up the "Create a New Macro" properties window. Type in "Hello" in the "Name of new macro" text box and hit OK (see Figure 7.28).

This creates a new Flowchart named "Hello" in addition to our main Flowchart. Enter into this Flowchart all the icons we used before to print out "Hello World," "Testing 1..2..3" (see Figure 7.29).

Once you have this information entered, use your mouse. Go to the upper left corner of the Hello Flowchart. Press the left mouse button and draw a rectangle around all the icons in the flowchart (see Figure 7.30).

Then release the mouse button. This will select all the icons between "BEGIN" and "END" (see Figure 7.31). Copy these icons with a Ctrl+C key combination.

Next create another macro (as before) and name it Where, see Figure 7.32.

Make sure the "Where" macro is selected; then hit Edit → Paste or Ctrl+V. This will paste the same icons from the "Hello" macro into the "Where" macro (see Figure 7.33).

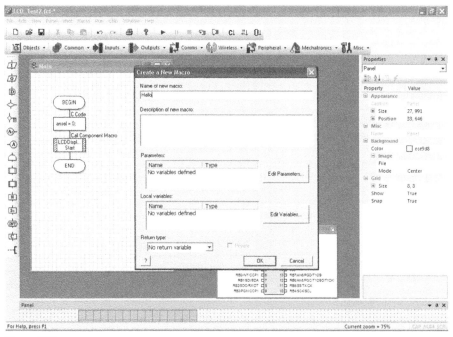

Figure 7.28

Next, double click on the first Call Component Macro for the LCDDisplay(0) and change the variable from "Hello World" to "Wherever you go" (see Figure 7.34).

Then double click on the second Call Component Macro for the LCDDisplay(0) and change the variable from Testing 1..2..3" to "There you are." Now you can reduce the physical size of the two macros "Hello" and "Where" that you created and put them on the side of your main Flowchart (see Figure 7.35).

We are keeping the macros shown on the screen as a reminder that they are there. We could also close the macros and remove them from the screen, and they would still be available for use from the main Flowchart. We created these new macros, now let's use them. Select and drag the Call Macro icon into the main Flowchart and release the icon under the LCDDisplay Start icon (see Figure 7.36).

Once the Call Macro icon is placed in the main Flowchart, double click on the icon to open its Properties window (see Figure 7.37).

In the Macro window of the Properties window, our two created macros Hello and Where are already listed. Select the "Hello" macro and click OK. At this point you can run the

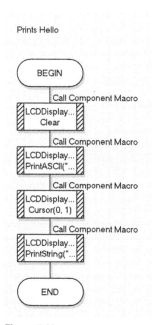

Figure 7.29

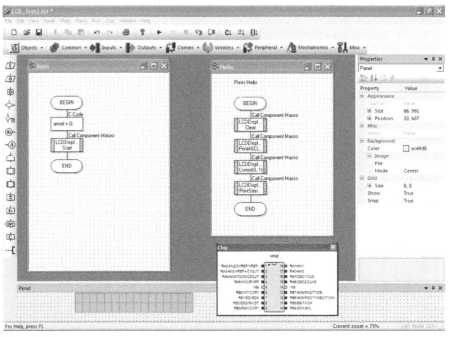

Figure 7.30

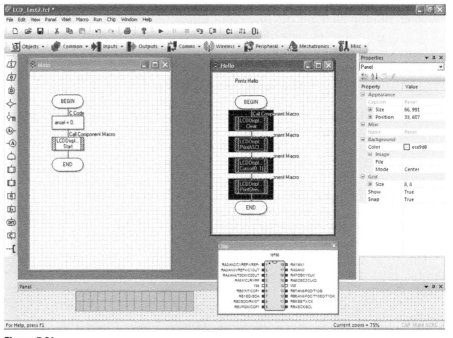

Figure 7.31

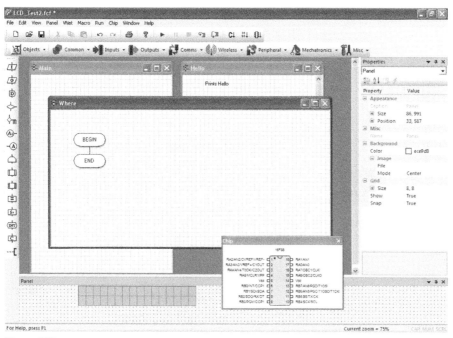

Figure 7.32

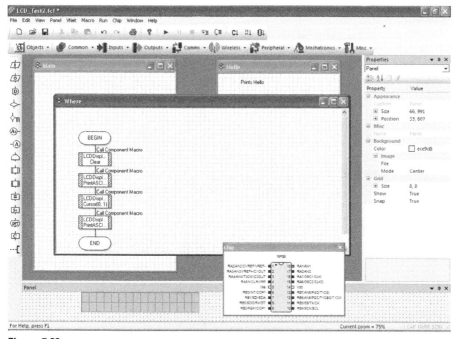

Figure 7.33

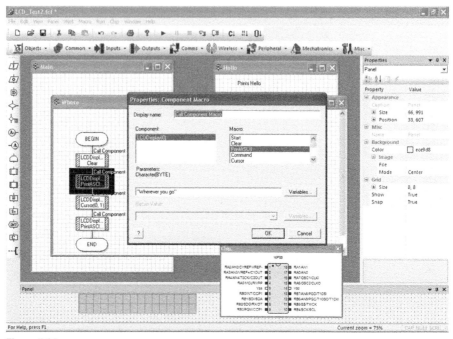

Figure 7.34

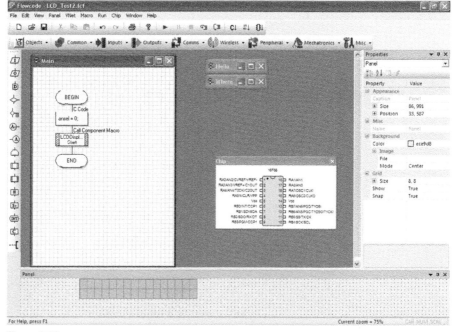

Figure 7.35

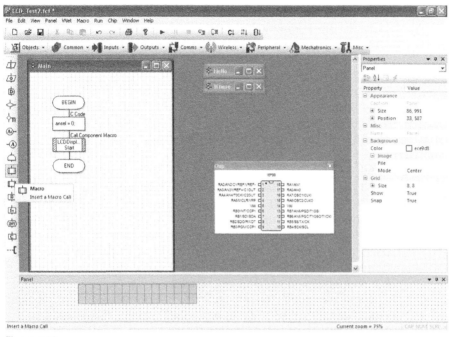

Figure 7.36

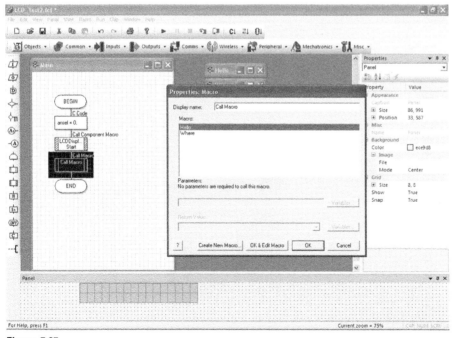

Figure 7.37

simulation and the text from the Hello macro is printed on the LCD. If the simulation doesn't work, double-check your work. Once the simulation works, add a 500 millisecond delay after this Call macro. Then add another Call Macro icon. Double click the icon to open its Properties window and select the Where macro.

Now run the simulation. The second macro will display half a second after the first. You can use this technique to populate your Flowchart with macro text and call it wherever and whenever it's needed from the main Flowchart program.

Using Strings

Flowcode provides for the use of string functions. Strings are byte arrays of ASCII characters and can therefore hold text and may be printed to the LCD. To print strings to the LCD you use the PrintString function. String macros may be created and used in a similar manner as just described for the PrintASCII function. Without repeating all the same material for strings, I'll go over what you need to get started.

First we need to put a String Manipulation icon in our Flowchart. Starting with a basic LCD main Flowchart for the 16F88, with the Code Component icon and LCDDisplay(0) Start macro, as shown in Figure 7.38, we add a String Manipulation icon.

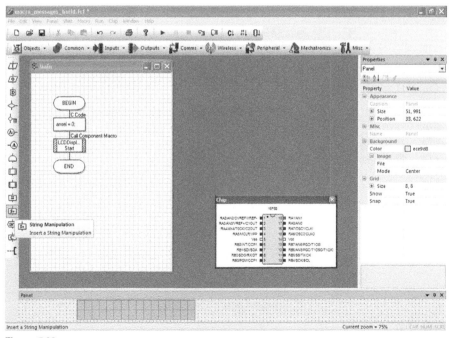

Figure 7.38

Double click on the String Manipulation icon to open its Properties window (see Figure 7.39).

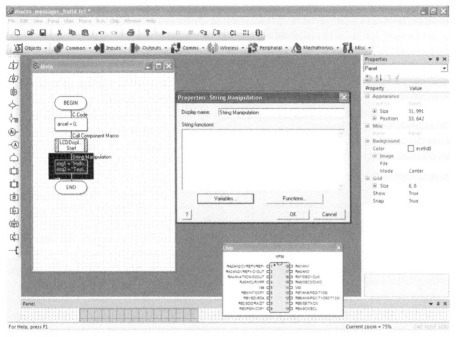

Figure 7.39

Next, click on Variables to open the Variable Manager window. Inside the Variable Manager window click the "Add New Variable" button to open the "Create New Variable" window and four strings; strg1, strg2, strg3 and strg4. Leave the default value of 20 bytes (see Figure 7.40).

When you have finished adding the string variable hit the Close button, which brings you back to the Properties window. In the String Functions text box, define the string functions with text as shown in Figure 7.41.

To use the string variable add a Call Component icon below the String Manipulation icon. Open the Properties window of the Call Component icon, select the PrintString macro and type in strg1 in the variables text box (see Figure 7.42). Hit OK and run simulation.

The program will print "Hello World" on the first line of the LCD. All the previous examples that we have run through using PrintASCII will work with PrintString. You can create text macros, as we have before using PrintASCII, and call them when needed from the main Flowchart.

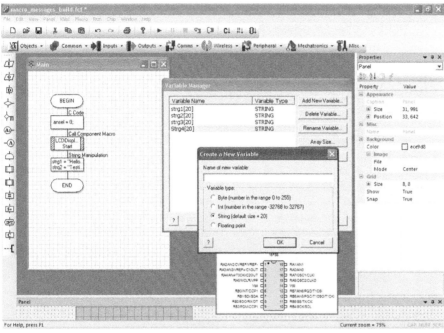

Figure 7.40

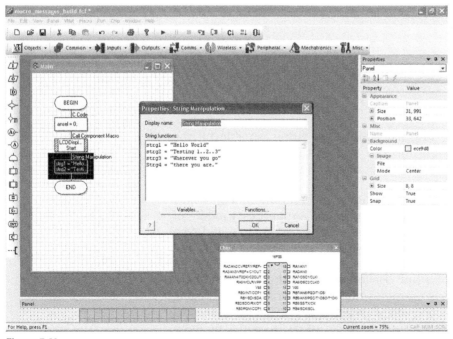

Figure 7.41

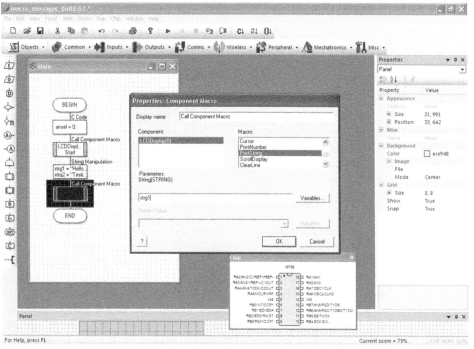

Figure 7.42

Creating Custom LCD Characters

I will complete this chapter with creating custom LCD characters. Custom characters will not show in simulation. Therefore you must program and upload the Hex file into the PIC microcontroller to see the results.

The LCD has eight unused characters in RAM (Random Access Memory) that the user may define and use. Each character is defined by a 5×8 pixel pattern. The standard character generator of the HD44780 holds the 5×8 pixel patterns for its standard ASCII characters in ROM (Read Only Memory) and cannot be changed. Each custom character is defined by a 5×8 pixel pattern. The pattern consists of five pixels across each row, eight rows high. Each 5 pixel row is represented by one byte (see Figure 7.43).

Each box in the bitmap has a value listed at the top of its column. When the box is colored in, the value is added to all the colored boxes across each row and placed in the total value column to the right. These total values are the byte values you will use to define your custom characters. Each character requires

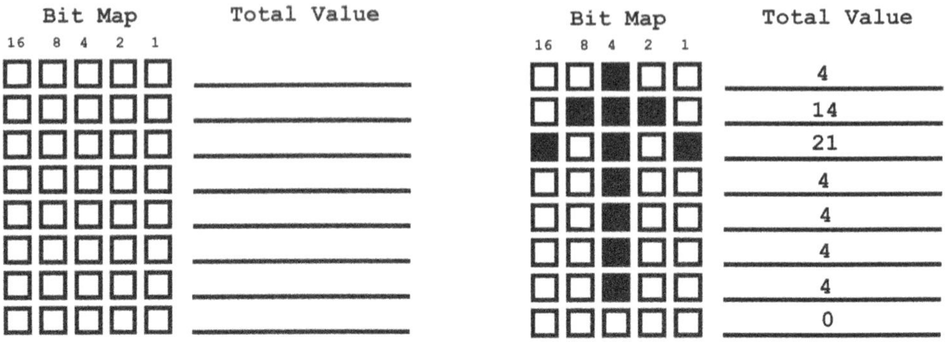

Figure 7.43

eight bytes of information to be defined. An up-pointing arrow character is shown as an example. A worksheet is provided at the end of this chapter to help you to define the eight custom characters. You do not need to define all eight characters; you can define just one character if that is all you need.

Flowcode provides a macro under the LCDDisplay(0) component to write to the character generator (CG) Ram of the LCD. Start with a basic LCD main Flowchart for the 16F88, with the Code Component icon and LCDDisplay(0) Start Macro, as shown in Figure 7.44. To this flowchart we add the LCDDisplay (0) macro RAM_Write. In the variables text box add the following data for the up-arrow; 0, 4, 14, 21, 4, 4, 4, 4, 0 and then hit OK. The first number "0" is the first character position in CG Ram where Flowcode starts writing the byte data.

To print out the custom character on the LCD, use the Print ASCII macro using the number as the character parameter, in this case 0 (see Figure 7.45).

NOTE: This procedure does not simulate in the Flowcode simulator; save the program, generate a Hex file and upload firmware into PIC microcontroller to check the custom character program.

If you ran your program and the LCD printed the up-arrow in the first position on the first line we can program and display all eight custom characters. Next, using Figure 7.44 as a guide, we will add several more custom characters. But we will put all of our custom character definitions inside its own macro Flowchart. Create a new macro Flowchart as shown in Figures 7.27 to 7.29 previously. Name this macro "Custom_Characters." For each set of character data listed below, create a Call Component icon on the Custom_Characters Flowchart, open the Call Components Properties window, select RAM_Write Macro and enter the

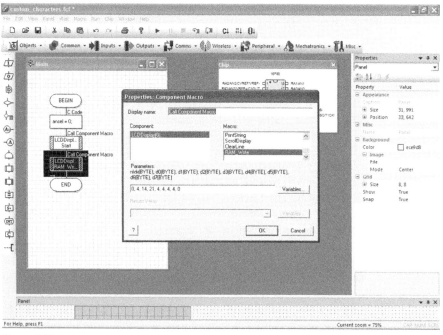

Figure 7.44

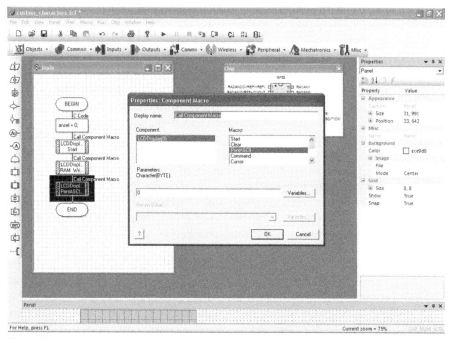

Figure 7.45

character data. At this time also change the Display Name of these icons to the name of the custom character being defined. Having unique names for Call Components will help to keep them all straight.

Character 0 "Up Arrow"; 0, 4, 14, 21, 4, 4, 4, 4, 0 as per the example given in Figure 7.44

Character 1 "Down Arrow"; 1, 0, 4, 4, 4, 4, 21, 14, 4

Character 2 "Upside Down Question Mark"; 2, 4, 0, 4, 4, 8, 17, 17, 14

Character 3 "Cross"; 3, 4, 4, 31, 4, 4, 4, 4, 4

Character 4 "Music Note"; 4, 3, 2, 2, 2, 2, 14, 30, 14

Character 5 "Locked"; 5, 14, 17, 17, 31, 27, 27, 31, 0

Character 6 "Un-Locked"; 6, 14, 16, 16, 31, 27, 27, 31, 0

Character 7 "MicroSoft Arrow"; 7, 30, 28, 28, 18, 1, 0, 0, 0.

Once this information is entered, the macro Flowchart should look like Figure 7.46 and we go back to the main Flowchart.

Look at Figure 7.47 for a reference as you read through the following instructions.

Add a Call Macro icon on the Flowchart beneath the LCD Start; choose Custom_Character as the macro to call. Next, add a Call Component icon under the Call Macro icon. Choose LCDDisplay(0) and choose the PrintASCII macro. Type in the Character text box "Custom Character." Add a Call Component icon underneath this icon, choose LCDDisplay(0) and choose the Cursor macro and type in 0,1 in the variables text box.

Add a Calculation icon underneath this icon. Define a new variable, call it x; make the variable type byte. Type in "x = 0" in the Calculation icon. Under the Calculation icon place an A: Connection point. Under the Connection icon add a Call Component icon. Choose LCDDisplay(0) and choose the PrintASCII macro. Type in the Character text box x with NO quotation marks.

Under this Call Component icon add another Call Component icon. Choose LCDDisplay(0) and choose the PrintASCII macro. Type in the Character textbox " ". That is a one character space inside quotation marks. This will print a space between the custom characters to make them easier to read.

Add a Calculation icon underneath this icon. Type in "x = x+1" in the calculation icon. Next add a Decision icon underneath this icon. Open its Properties window; select the "Swap Yes and No" option. Type in x = 7 in the If text box. Then hit OK. In the NO leg of the decision icon, add the A jump point. Save the file.

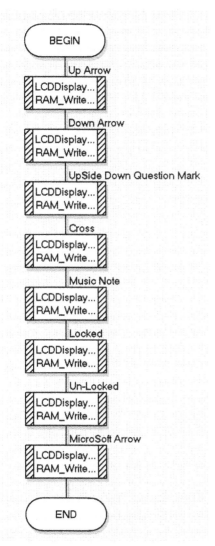

Figure 7.46

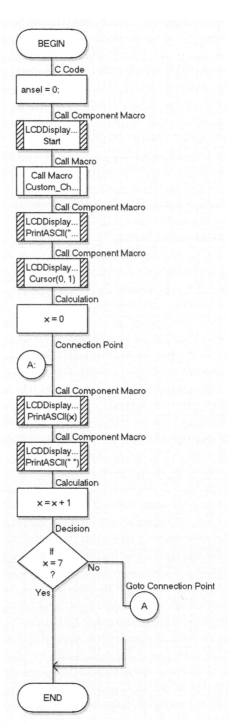

Figure 7.47

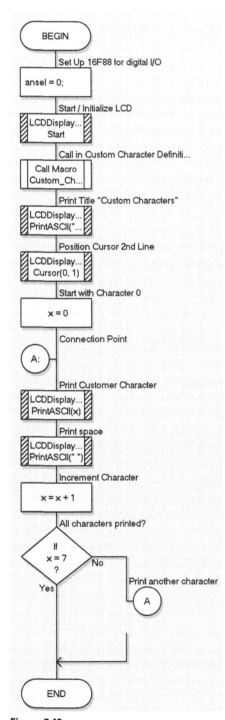

Figure 7.48

You cannot see this Flowchart work in simulation. Custom characters are not supported in simulation. Before we test this program, let's go back and provide names for our Flowchart icons that give an idea of their function. This is a good practice to get into. It will make reading and understanding the Flowcharts much easier after a few months have passed, when what seems so obvious and clear to you now becomes only a vague recollection.

Look at Figure 7.48. This shows the identical Flowchart to that in Figure 7.47, with the icons renamed. This Flowchart will be easier to understand a year from now when you are trying to figure out how you did something.

Compile the program to a Hex file. Load the Hex file into your programmer. If you are using the EPIC programmer check the settings (see Figure 7.18). Without the proper settings the program will not function. Upload your Hex file (firmware) into the 16F88. Place the microcontroller back into its circuit on the solderless breadboard and apply power. If you did everything right, the display on the LCD will appear as in Figure 7.49.

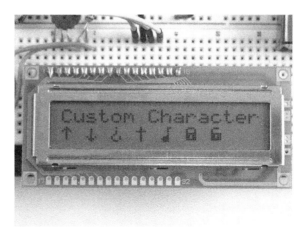

Figure 7.49

Custom Character Worksheet

See Figure 7.50.

Custom Character Worksheet

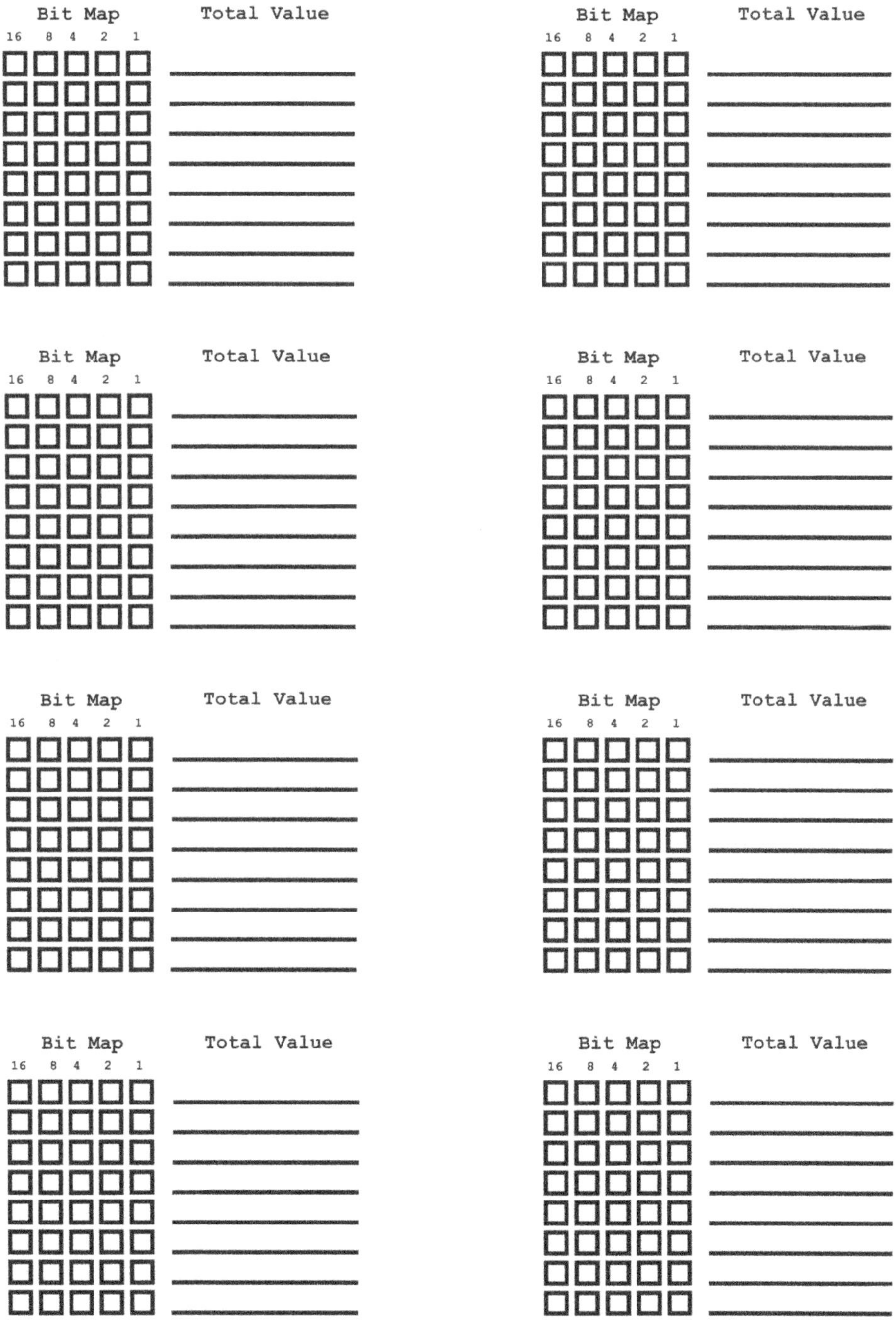

Bit Map	Total Value
16 8 4 2 1	

Bit Map	Total Value
16 8 4 2 1	

Bit Map	Total Value
16 8 4 2 1	

Bit Map	Total Value
16 8 4 2 1	

Bit Map	Total Value
16 8 4 2 1	

Bit Map	Total Value
16 8 4 2 1	

Bit Map	Total Value
16 8 4 2 1	

Bit Map	Total Value
16 8 4 2 1	

Figure 7.50

8

RS232 SERIAL COMMUNICATION

Flowcode has a number of communication protocols (see Figure 8.1).

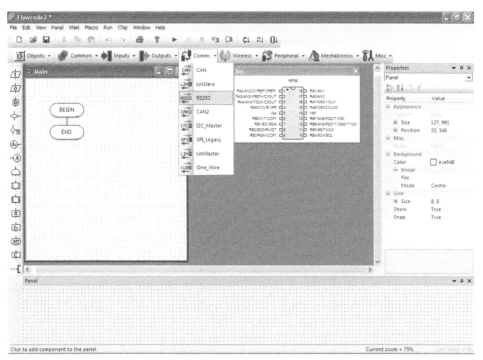

Figure 8.1

In this chapter we will work with the RS232 protocol. This Flowcode protocol performs serial, asynchronous communication. The term asynchronous means "without a clock" or "out of sync," and this is the communication we want. Let me explain.

PIC Projects for Non-Programmers. DOI: 10.1016/B978-1-85617-603-3.00008-7

Synchronous communication requires a shared clock line between the transmitter and the receiver. The clock line signals when data on the serial line are valid. Asynchronous communication does not have a clock line. In lieu of a shared clock line, this protocol demands strict time and framing control of each byte transmitted and of each bit within that byte. In this way the time elapsed, beginning with the start bit, becomes the sync factor and determines when the data on the line are considered valid. Let us begin by looking at a diagram of our serial signal (see Figure 8.2).

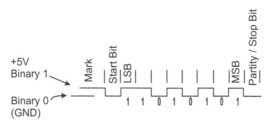

Figure 8.2

The first thing to notice on the diagram is the voltages used on our serial line are +5 volts to 0 volts (GND). These voltages are PIC microcontroller-compatible. True RS232 voltages swing from positive to negative voltages and can vary by up to ±12 volts. We will not be working with RS232 voltages, only RS232 signals.

Asynchronous communication requires the use of a start bit and a stop bit. The serial line is held in a "Mark" condition, which means the line being held high (+5 volts). Mark may also be considered binary "1." The start bit always precedes the byte being transmitted and brings the serial line to ground momentarily. Bringing the line to ground is a binary "0" and may be referred to as a "Space." The start bit wakes up the receiver and starts the timing cycle that will keep all the following bits in sync and properly received.

In our communication protocol, the bit information that follows the start bit is an 8-bit byte followed by a stop bit. In this communication the Least Significant Bit (LSB) is sent first, followed by the Most Significant Bit (MSB). This is not the only serial protocol available. Other serial protocols are possible, for instance, the data being transmitted could also be 7-bit ASCII; there could be an error-checking parity bit and possibly two stop bits. Therefore our standard data package consists of 10 bits: our 8-bit byte plus a start and stop bit. Our first program uses a 1200 baud (bits per second) speed. So 120 bytes of data are transmitted each second.

The timing in asynchronous communication must be strictly followed. Without a clock line, elapsed time is used to determine when the value on the serial line is valid. For instance, at 2400 baud each bit is on the line for 0.4 milliseconds, and must be seen and captured by the receiver in that time period. So, in order for everything to work properly, the transmitter and receiver cannot vary by more than 5% from the ideal baud rate. If the baud rate does vary more than that, then over the course of the ten bits needed to transmit a byte, the last bit may fall out of sync, may not be valid and may corrupt the entire byte.

Error Detection Algorithms

Error detection algorithms that detect sync errors are common in most communication packages. However, Flowcode's PIC serial communication does not, as of this writing, contain error detection algorithms. In fairness to Flowcode, I have yet to see any error detection algorithms in other PIC compilers I've worked with.

Parity

Parity uses the stop bit for error detection. Parity may be odd, even, or as in our case none (no error detection). Just because we're not using it doesn't mean we shouldn't understand it. Looking back at Figure 8.2 we see that serial transmission consists of a bunch of binary 1s and 0s. Using even parity, the receiver would add up all the binary 1s. If the number of bits transmitted equaled an odd number it would expect the stop bit to be "1," to make the number of binary 1s even. If the number of bits transmitted equaled an even number it would expect the stop bit to be "0," keeping the number of binary 1s even. When parity is set to even and an odd number of bits are received, this is called a parity error, and the entire byte is discarded. The same procedure is used for odd parity, with the parity bit set to keep the count of binary 1s to an odd number.

Baud Rates

Flowcode allows one to choose a variety of baud rates: 1200, 2400, 4800, 9600, 19200 and 38400. Data is sent as one start bit, eight data bits, no parity and one stop bit. I would not recommend using the higher baud rates that are available unless you are also using a fast crystal (Xtal). Standard crystal speed is 4 MHz. To use the higher baud rates, choose a 20 MHz crystal and set the crystal speed in the Project Options window, discussed in the next paragraph.

Setting Clock Speed

When I started using the RS232 component I could not get the communication to work properly. I had my crystal set, or so I thought, so I started looking for all sorts of complex reasons why communication was not occurring – start bits, stop bits and all the bits in between. When I exhausted all the troubleshooting problems I could think of I contacted Flowcode's technical support at Matrix Multimedia.

After a few emails to Matrix Multimedia technical support I was later advised to check the speed and timing of the microcontroller. The test was to blink an LED on and off for 1-second intervals. When I performed the test, the LED blinked at 5-second intervals. The microcontroller program was somehow set at five times the speed it was actually running at. Now I was getting somewhere. I had to set the Xtal to XT in the Chip configuration screen, go to menu item Chip, and then Configure (see Figure 8.3). In other compilers I have used XT

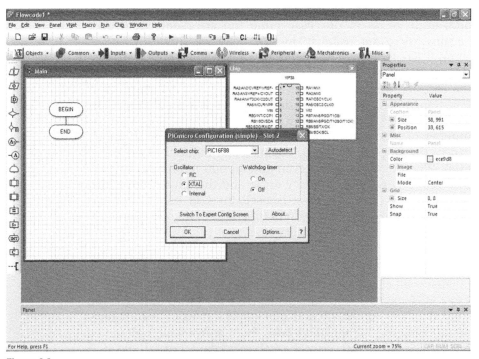

Figure 8.3

defaults to a 4 MHz Xtal, and I assumed Flowcode did the same. I was incorrect in that assumption.

Flowcode has another window called Program Options. Go to menu item View and then Project Options (see Figure 8.4).

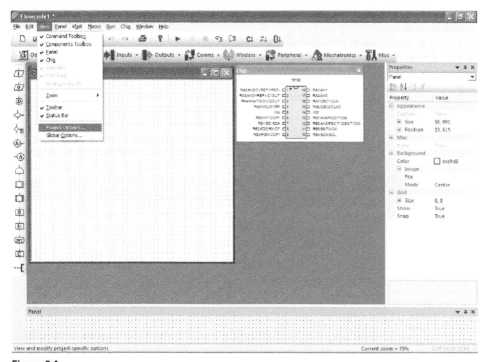

Figure 8.4

Selecting this menu option opens the Project Options window. In this window we can properly select the clock speed of our circuit. I selected 4,000,000, which is 4 MHz, and hit OK (see Figure 8.5).

Once this change was made, communication proceeded without a problem. Since you are starting with this correctly setup you will not have any problems. Let's proceed.

Serial LCD

The device we will connect to in this chapter is a serial LCD. This LCD is similar to the device we used in the last chapter. Now you may ask "Why do another LCD display?" Well, we are using it to describe Flowcode PIC RS232 serial communication, but other than that there are different applications where one type of LCD display is better than the other.

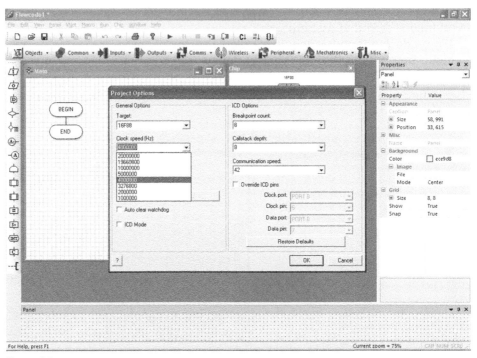

Figure 8.5

Applications

First there's the difference in cost. An LCD module discussed in the last chapter can be purchased for $7.95 for a single unit. In quantity the price is lower. A serial LCD we are discussing in this chapter costs $22.95. If you are creating a commercial product that will be manufactured and sold, using an LCD module is the least expensive way to add an LCD to your project. On the other hand, the serial LCD only requires a single I/O pin for the serial line. If you are short on I/O pins on your microcontroller project, the serial LCD might be the way to go. The serial LCD does require +5 volts of regulated power and must share a common ground with the controlling PIC microcontroller for the serial line to work.

Serial LCD Power Supply

The serial LCD power supply must be regulated to 5 volts. Any voltage that is greater than 5.5 volts will burn out the microcontroller and possibly the LCD module itself. The serial LCD is easier to use and has fewer connection wires (see Figure 8.6).

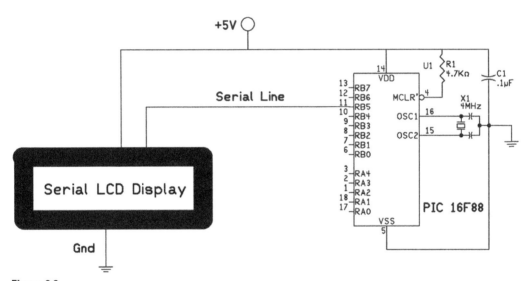

Figure 8.6

Our LCD display requires just three wires: +5 volts power, ground (GND) and the serial line itself. The microcontroller and LCD must share a ground in order for the serial line to work properly. The LCD display can be set to a variety of baud rates. A 4-position DIP switch on the back of the serial LCD is used to select various baud rates. When the switch is in the "up" position (open) its value is "1." When the switch is in the "down" (closed) position value its is "0." Baud rate code encoding of the DIP switch is only checked upon LCD start-up. Changing baud rate switches while the LCD is powered on will not change baud rate. To change baud rate, power down the LCD; then change switches and power up the LCD display.

SR	Switch 4	Switch 3	Switch 2	Switch 1	Baud Rate
1	0	0	0	0	TRUE 300
2	0	0	0	1	TRUE 1200
3	0	0	1	0	TRUE 2400
4	0	0	1	1	TRUE 4800
5	0	1	0	0	TRUE 9600
6	0	1	0	1	TRUE 19 200
7	0	1	1	0	TRUE 38 400
8	0	1	1	1	TRUE 57 600
9	1	0	0	0	INVERTED 300
10	1	0	0	1	INVERTED 300
11	1	0	1	0	INVERTED 19 200
12	1	0	1	1	INVERTED 9600
13	1	1	0	0	INVERTED 4800
14	1	1	0	1	INVERTED 2400
15	1	1	1	0	INVERTED 1200
16	1	1	1	1	INVERTED 300

Self-Test Mode

It's important to wait for two seconds on power-up before sending any data or command to the LCD, due to the initialization period. For normal serial operation the serial input pin on the LCD should be held at mark level (i.e., High in True Mode and Low in Inverted Mode) on power-up. If the lines are not held at these values on start-up, the LCD will go into a self-test mode.

The self-test mode shows the current baud rate on the LCD. This is a simple way to check the baud rate that the LCD is set to. Once in self-test mode, to continue the normal LCD display functions module, invert the serial line value. This will reset the LCD for normal operation. So if the serial input pin is kept at non-mark level (Low in True Mode and High in Inverted Mode) on power-up, the LCD module will enter a self-test mode, where it will show current baud rate selected. To print characters on LCD, ASCII representation of characters should be sent serially at selected baud rates as shown in our first program.

First Program PRG 8.1

Our first program will print out the message "Hello World" on the LCD. The program is set to transmit serial data at 1200 baud, in True Mode. Therefore we must set our serial LCD display to receive data at 1200 baud in True Mode. Start a new Flowcode chart; select COMMS > RS232 (see Figure 8.7).

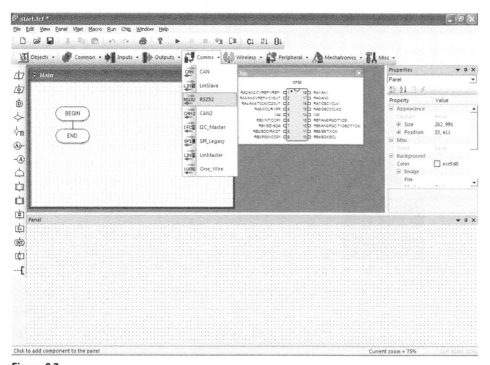

Figure 8.7

This action places the RS232 component in the Panel section (see Figure 8.8).

The first thing we see is that the RS232 component has three windows: Characters Sent, Characters Received and Characters in Queue. We will be focusing on transmitting characters to our serial LCD. Select the RS232 components and click on the Ext Properties, to bring up the "Edit Component Properties" window (see Figure 8.9).

With this Properties window open, set the baud rate to 1200. Under the Tx/Rx label select the UART option. UART is an acronym for "universal asynchronous receiver/transmitter." It is hardware built into the PIC microcontroller specifically for serial communication. Leave the other options at their default settings and hit OK. The UART option in the Flowcode compiler allows you to use the built-in UART inside the PIC 16F88 microcontroller. When using the UART for serial communication the RB5 pin is automatically configured to transmit the serial data.

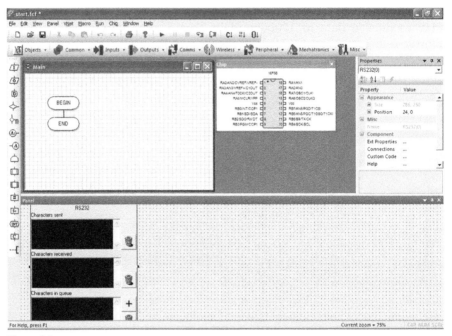

Figure 8.8

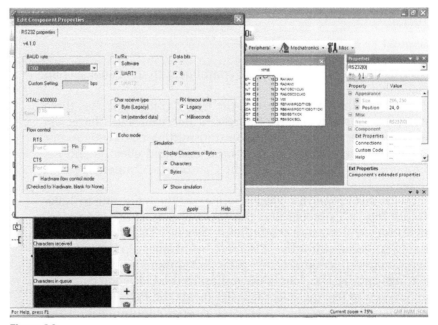

Figure 8.9

While we are not using it in this application, RB2 is automatically configured to receive data. Build the circuit schematic illustrated in Figure 8.6.

First Serial Program

The first icon we will add to our program is a two second delay. Upon power-up, the LCD may require up to two seconds for its own initialization. Label the Delay icon "LCD Initialization."

The next icon to add is the Component. Place the Component under the Delay icon. Double click to open up its Properties window. Select the RS232(0) component and then select the "SendRS232String" macro. In the "Variables" text box enter " Hello World." (see Figure 8.10). Make sure to add a blank space before the "H" in Hello. Often it takes the serial character transmitted to the LCD to wake it up after initialization. If you don't add the blank space, you may lose the "H" in hello. Label this Component Macro "Send message."

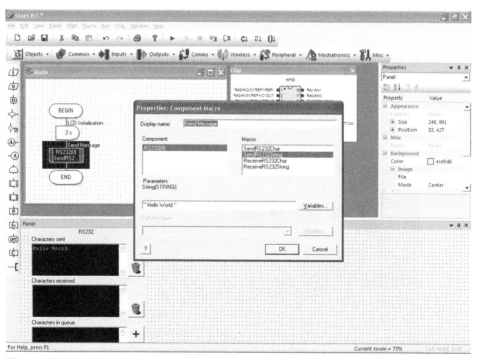

Figure 8.10

Save the program and compile to a Hex file. You can simulate the program and see the message "Hello World" printed in the Characters Sent window of the RS232 component. When programming the PIC microcontroller, make sure you have the 16f88 chip selected and the values in the MeProg – Configuration as shown in Figure 8.11.

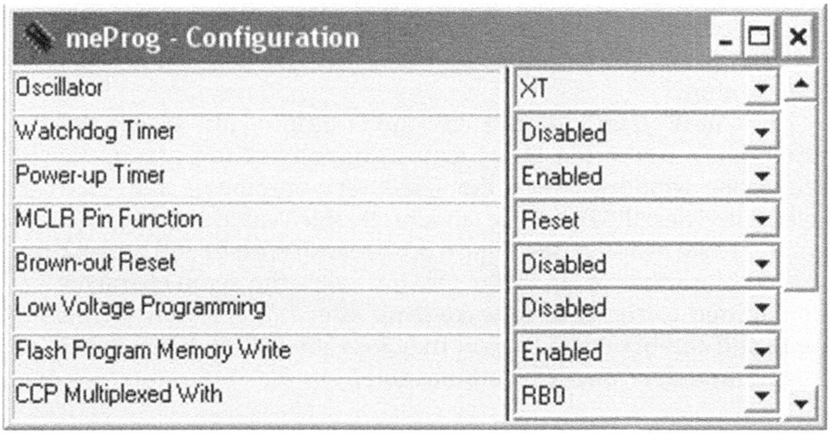

meProg - Configuration	
Oscillator	XT
Watchdog Timer	Disabled
Power-up Timer	Enabled
MCLR Pin Function	Reset
Brown-out Reset	Disabled
Low Voltage Programming	Disabled
Flash Program Memory Write	Enabled
CCP Multiplexed With	RB0

Figure 8.11

Figure 8.12 shows the circuit built on the solderless breadboard.

Figure 8.12

Using the serial LCD we have the same features and options that are available to us as we used for the parallel LCD interface in Chapter 7. The following table is a list of the popular commands available. To initiate a command, the command must be prefixed with the decimal number 254. The serial LCD will treat any number following a 254 as a command.

Instruction Codes for Serial LCD Display

Code	Instruction
1	Clear screen
2	Bring cursor to home position (top left)
8	Blank without clearing
12	Make cursor invisible/restore display if blanked
13	Turn on visible blinking cursor
14	Turn on visible underline cursor
16	Move cursor one character left
20	Move cursor one character right
24	Scroll display one character left
28	Scroll display one character right
192	Move cursor to first position on second line

To send a command to the serial interface requires the use of two Component Macros; each one sends a decimal number. The first number sent is 254, followed by the command number. We can data load custom LCD characters, scroll the display and use the off-screen memory. The next example prints a message to the second display line on the LCD, just to illustrate using commands. We will send the number 254, followed by 192 and then print the message.

Second Program 8.2

Starting with the Flowchart program shown in Figure 8.10, we add three Component icons above the "Send Message" icon. In the first Component icon we open up its Properties window, and select the RS232(0) component and the SendRS232String macro. In its variable text box we enter " ", a blank space enclosed in quotation marks. Enter a display name of "Wake up" and hit OK.

This Component Macro is necessary to begin serial communication with the LCD. Without this macro, the LCD will miss the prefix command. You only need to wake up the LCD once.

In the second Component icon we open up its Properties window, and select the RS232(0) component and the SendRS232Char macro. In the Variables text box enter the number 254, enter a display name of "Command Prefix" and hit OK (see Figure 8.13).

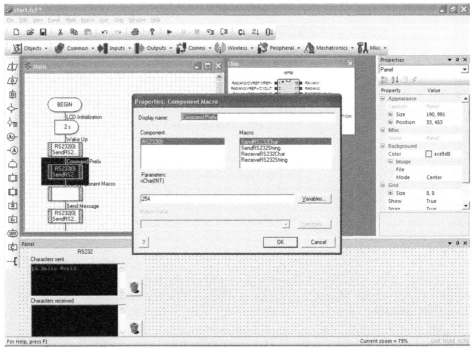

Figure 8.13

Next, open the second Component icon, open up its Properties window, and select the RS232(0) component and the SendRS232Char macro. In the Variables text box enter the number 192, enter a display name of "Command: Move cursor to first position on second line" and hit OK. Save the program, and compile and upload your firmware into the 16F88 PIC microcontroller. When the program is run in your circuit, "Hello World" will be displayed on the second line. If you like, you can remove the blank character space before the H in "Hello World." The function of that blank space has been taken over by the first RS232(0) component.

Software Control

In the last example we used the hardware UART in the PIC 16F88. It is also possible to use software control. Why use software RS232 control? Software allows you to choose which I/O pins on the PIC microcontroller will be used for transmitting and receiving serial data. The software control needs a faster crystal. The standard 4 MHz crystal used in the circuit was just a tad too slow and gave garbage characters on the LCD. I switched a 16 MHz crystal into the circuit and serial data transmitted by the PIC was received by the LCD properly. I think one could use an 8 or 10 MHz crystal, but if one wants to use faster than 1200 baud rates, you should opt for a 16 MHz or 20 MHz crystal.

Setting-up for Software Control

First switch the 4 MHz crystal in the circuit to a 16 MHz or 20 MHz crystal. In this example we are using a 16 MHz crystal. We can continue using the program from our last example; if you like, you can perform a "Save as" and save the program under a new name, such as PRGM_8_3.fcf. Select the Program options from the View menu (see Figure 8.14).

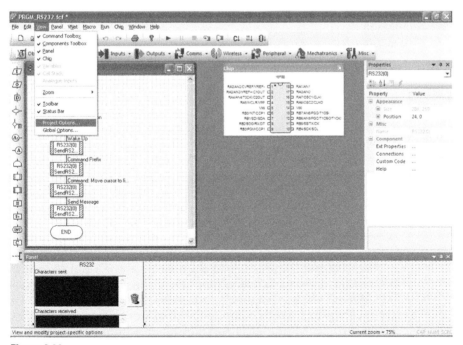

Figure 8.14

This opens the "Project Options" window (see Figure 8.15). In the "Clock Speed (Hz)" text box, type in 16,000,000 and hit OK. If you look at the pull-down menu for this item, 16,000,000 (16 MHz) is not a default option, but you still have the ability to write in your own clock speed.

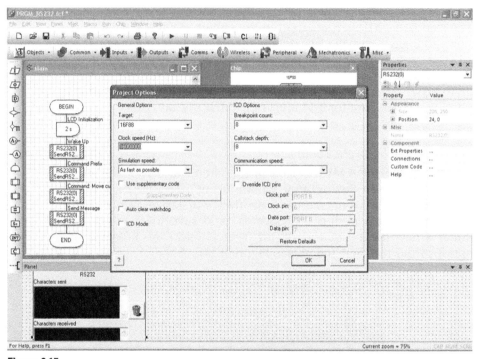

Figure 8.15

With the clock speed properly selected in the project options, select the RS232 control in the Panel portion of the screen. Then open up the Ext Properties window as shown in Figure 8.16.

Under the Tx/Rx options, select the Software option. Double check that the baud rate is still 1200 baud; if it has changed due to changing the crystal, pull down the options menu and reselect 1200 baud. Hit OK. Next, open up the Connections properties window, as shown in Figure 8.17.

Connect the RX pin to Port B bit 0 and the TX pin to Port B bit 1 and hit Done. While we are not using the RX pin, it must be defined anyway or being non-selected it would cause an error when we tried to compile the program to Hex. On the circuit board, move the LCD serial input pin from the 16F88 pin 11 (RB 5) to pin 7 (RB 1) (see Figure 8.18).

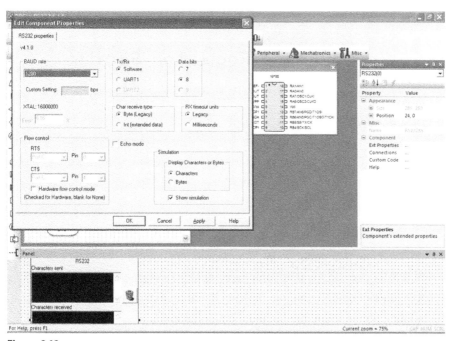

Figure 8.16

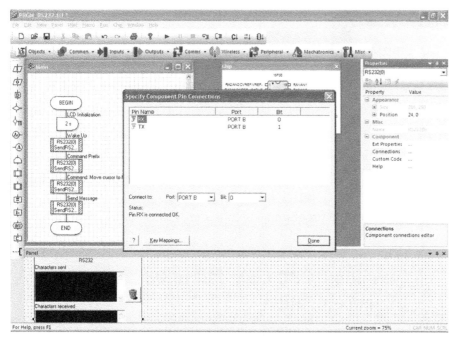

Figure 8.17

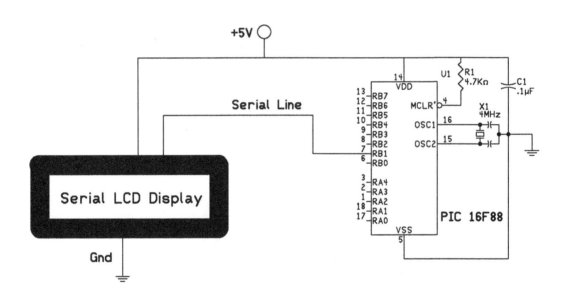

Figure 8.18

Run the program as before and "Hello World" will be printed on line 2 of the LCD display.

Creating a PC Link: PIC to PC Communication

The RS232 protocol allows your PIC to communicate to PCs using a simple PC terminal communication program like

"HyperTerminal." Newer PCs and laptops have discarded the DB9 (RS232) and may not have a serial port. Everything is migrating toward USB. Not to worry, there are USB-to-RS232 adaptors readily available that will allow you to keep simple serial communication between your PC and PIC.

USB or Serial?

While my PC does have an RS232 serial port, I opted to use a USB-to-serial adaptor to illustrate the procedure, as this is more likely to be the case you will encounter. If your computer has a serial port, you can skip the USB adaptor section. Keep one thing in mind; it is essential that you can communicate between your PC and your PIC using either an RS232 (COM) port or through the USB port, with an USB/RS232 adaptor.

USB-to-RS232 Adaptor

There are many USB-to-RS232 adaptors on the market. Figure 8.19 shows a photograph of the USB-to-serial adaptor I used. There is nothing special about this adaptor, and its usage should be common to most other USB-to-RS232 adaptors. On one end of the adaptor is a standard USB plug. This end plugs into your computer. The opposite end has a DB9 connector. The DB9 connector is where we will attach our circuit to communicate with the PC.

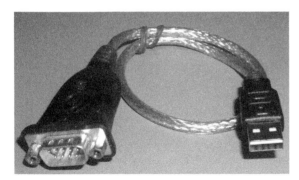

Figure 8.19

This USB-to-serial adaptor comes with an installation disk. When the software is installed it will create a pseudo COM (communication) port on your computer. The COM port is the RS232 serial port and will be available when you have the USB adaptor plugged into your computer's USB port. Computers may have more than one COM port. They could have a dozen if required. Therefore COM ports are labeled with a number to identify individual ports, such as: COM1, COM2, COM3. My computer had a few COM ports (COM1, COM 4 and COM 5) so when adding the USB-to-serial adaptor, the installation software created a COM 3 port for using the USB adaptor, as COM 3 was an unassigned COM port number at the time of installation. However, the installation software may choose another COM port number and not one necessarily in any specific order.

Finding the USB COM Port

If you are like me, it's easy to forget which COM port is assigned to the USB adaptor. One quick way to see if your computer has COM ports or to check the number of the COM port assigned to the USB adaptor is to follow this procedure. Go to the Start menu button, then to Settings and then to Control Panel. Open up the Control Panel. Once the Control Panel window opens find the System icon and double click on it to open this window (see Figure 8.20).

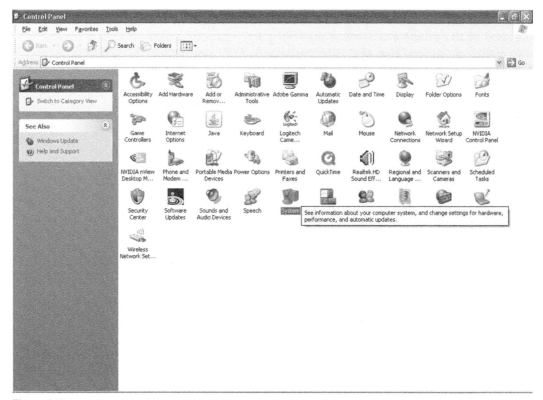

Figure 8.20

When the System Properties window opens, select the Hardware tab, then click on and select the "Device Manager." When the Device Manager opens look for the "Ports (COM & LPT) line. Click on the "+" sign to expand the item. In the expanded lines you will see the COM port assigned to the USB adaptor and highlighted in Figure 8.21.

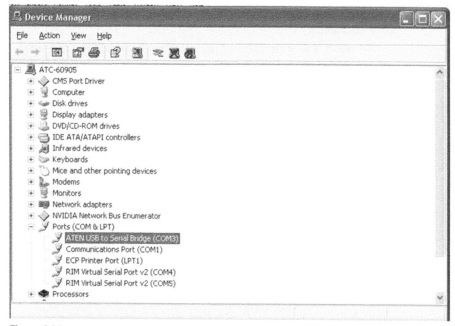

Figure 8.21

My USB adaptor is assigned to COM3. In order to see the USB adaptor's COM port, the USB adaptor must be plugged into the computer's USB port. If it is not, the COM port will not be seen in the Device Manager. Once you know which COM port is assigned you can close out the Device Manager's window and the Control Panel.

Changing the COM Port Setting on the USB-to-RS232 Adaptor

Suppose the adaptor program assigned the USB to a high COM port number like COM11 and let's further assume your software can't see any COM port above COM4; what do you do? Double click on the USB listing in the Device Manager to open up its Properties window. The second tab in the Properties window should be Port Settings. Click on Port Settings to open up that tab. At the bottom of this tab screen is a button titled Advance. Click on the Advance button. This opens another screen that has a pull-down menu where you can select any open COM port for your USB device. Select the COM port you want to use and hit OK. Exit all the way out of the control panel.

DB9 Connector

We need a female DB9 connector to connect the USB adaptor to our circuit (see Figure 8.22). This DB9 connector is a right-angled PC mount connector. In the DB9 description, the letters PC mean printed circuit (board).

Figure 8.22

To communicate with the personal computer (PC) we only need to connect our PIC to three lines of the DB9 connector. Pin 2 is the RX (Receive pin), Pin 3 is the TX (Transmit pin) and Pin 5 is ground. To simplify connecting our DB9 connector I modified it. First I removed all pins except pins 2, 3 and 5. Then I straightened the remaining pins so I could plug those pins (and the DB9 connector) into our solderless breadboard (see Figure 8.23).

Figure 8.23

If Flowcode had the ability to invert its serial signal we could have directly connected the PIC microcontroller to the DB9 connector and PC using two resistors. Unfortunately this is not the case. Therefore we must employ a dual line driver/receiver MAX232 IC. In addition the MAX232 properly matches the PICs TTL +5 volt logic signals to the ± voltages used on EIA232 communications. The MAX232 integrated circuit supplies standard EIA232 voltages from a +5 volt source. The schematic is shown in Figure 8.24.

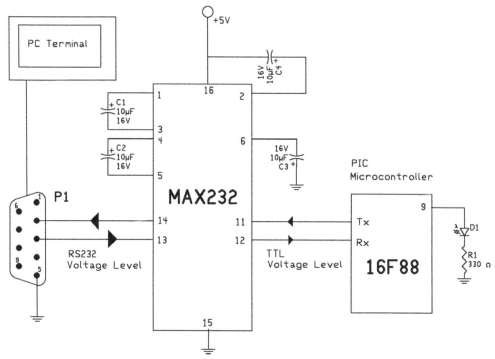

Figure 8.24

The schematic in Figure 8.24 focuses on the MAX232 IC and the components needed for it to be incorporated into our circuit on the solderless breadboard. Figure 8.25 shows the circuit built on our solderless breadboard. Don't forget the D1 LED and 330 ohm resistor connected to pin 9 on the PIC microcontroller. The usefulness of this LED will be explained a little later.

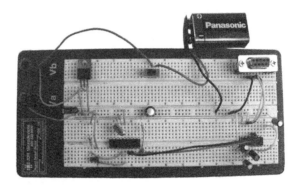

Figure 8.25

Our first PC terminal program will have the PIC send data to the PC. The PC will display the information on its screen.

HyperTerminal/PuTTY

HyperTerminal is free telecommunications software that has always been included with the Windows operating system. The latest versions of Windows, Vista and WIN7 no longer have this program. For those users I recommend a free Telenet program called PuTTY. I will give directions for both, starting with HyperTerminal.

If you have HyperTerminal or are using PuTTY, both use the following serial communication protocols listed. We must set up our communication parameters to match the serial parameters of our PIC microcontroller. This is 1200 baud, 8-bit, no parity and one stop bit. Before you start HyperTerminal make sure your COM port is available. If you are using a USB adaptor make sure it's plugged in first, so you can connect to the COM Port.

To start HyperTerminal go to your Start button → Programs → Accessories → Communications → HyperTerminal.

If this is the first time HyperTerminal has ever been run on your computer, the start screen will ask for a location and area code. Fill in this information and click OK to get to the communication screens. The first communication screen "Connection Description" will ask for the connection name. I fill in the textbox with the name Flowcode and hit OK (see Figure 8.26).

Figure 8.26

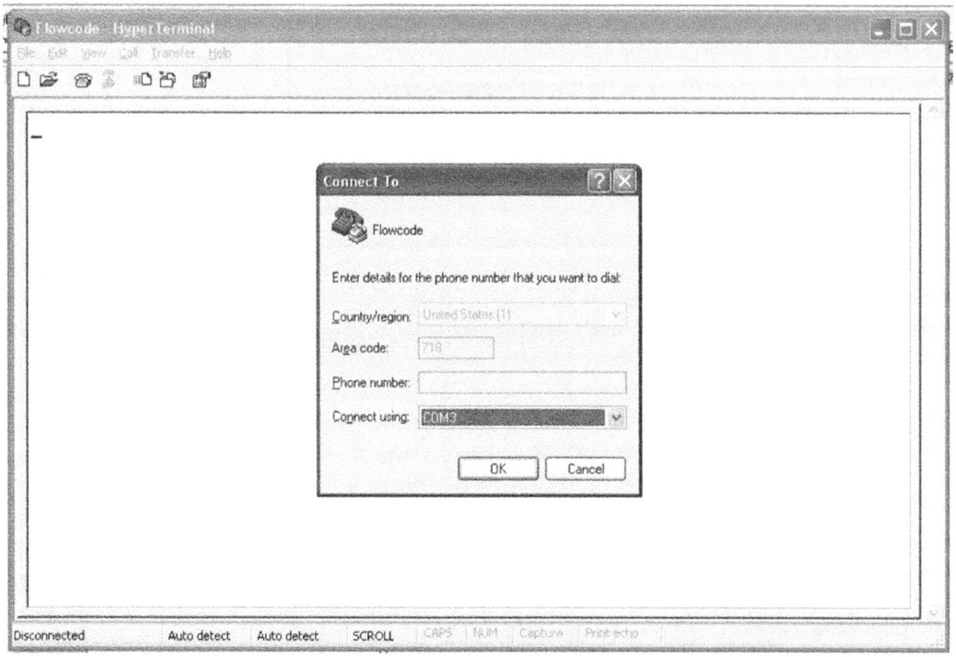

Figure 8.27

The next screen asks for "Connect Using:" . I choose COM3 as my port, and hit OK (see Figure 8.27).

The next screen asks for the communication properties of the COM3 port. Choose the following properties:

Bits per second = 1200

Data bits = 8

Parity = None

Stop bits = 1

Flow control = None (see Figure 8.28).

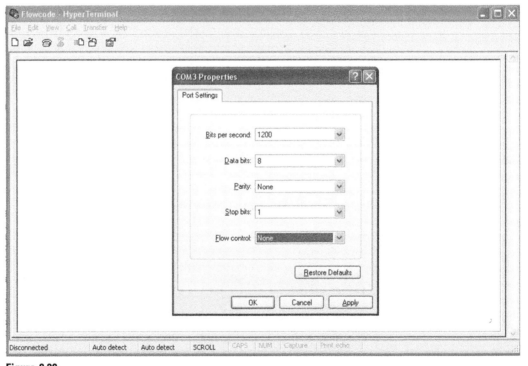

Figure 8.28

Next click OK. This opens up the HyperTerminal communication screen (see Figure 8.29).

Notice that the telephone icon has the telephone receiver off the hook, meaning it's connected to the COM3 port. If you are running with HyperTerminal you are set to program your PIC to communicate to the PC. Skip the PuTTY section that follows and pick up again with "Programming for PC Communication."

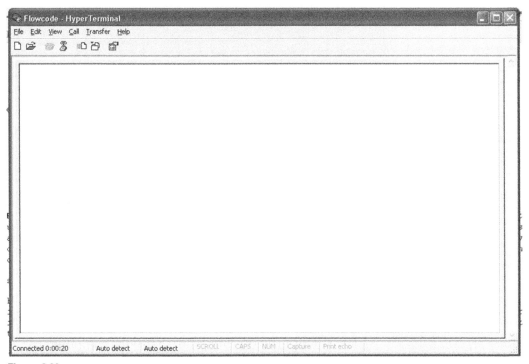

Figure 8.29

PuTTY

Do an Internet search for PuTTY and download the software. I downloaded the Putty.exe from: http://www.chiark.green-end.org.uk/~sgtatham/putty/download. html. Run the putty.exe file from your hard drive. When the software is run, the PuTTY configuration screen opens. First set the "Communication Type" to serial by selecting that option. Then I set my serial line to COM3 and my speed to 1200 (see Figure 8.30).

Click the button labeled "Open" which opens a PuTTY communication screen (see Figure 8.31).

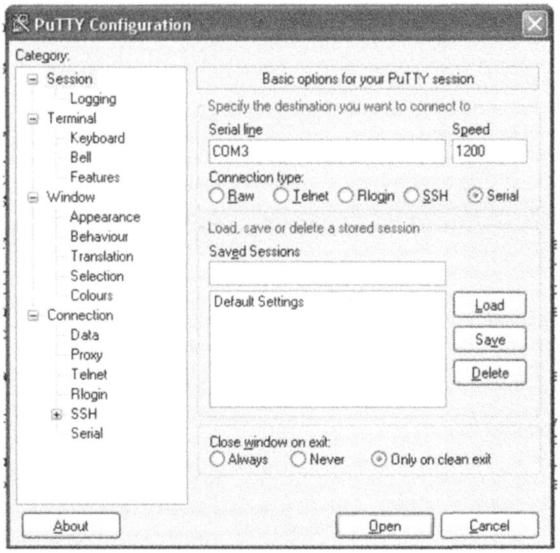

Figure 8.30

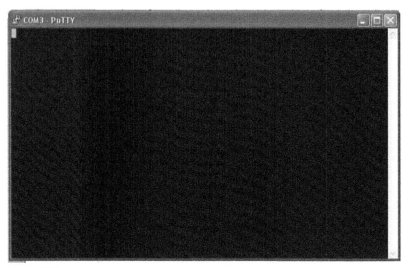

Figure 8.31

Programming for PC Communication, PRG8.4

Serial communication with your personal computer has a different set of commands available than with the LCD. This is the ASCII command character set. There are many commands available; we will use two common commands in the following example. The two commands are Carriage Return (CR) and Line Feed (LF). These two commands will allow us to enter a new line of text, after we have already printed a line of text to the PC terminal.

Carriage Return "Chr$(13)" brings the cursor back to the first text position on the line. Line Feed "Chr$(10)" drops the cursor down to the next line. So these two commands used in unison bring the cursor to the first position on the next printing line. Save your existing 8.3 program as 8.4. We begin by modifying our existing LCD Flowcode program. Change the time elapse in the first Delay icon from two seconds to one second. Change the name of the Delay from LCD Initialization to "Initialization." Next remove the next three Component icons titled "Wake up," "Command Prefix" and "Command: Move Cursor." Leave the Send Message component icon command as it is.

Next add a new Component icon after the "Send Message" icon; title this icon "Carriage Return." Open up the icon's Properties window (see Figure 8.32). Select the RS232(0) component; then select the "SendRS232Char" in the macro window. In the Parameters text line enter the number 13 and hit OK.

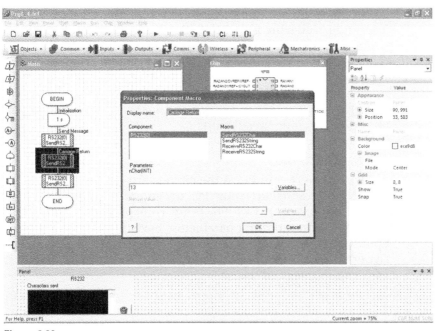

Figure 8.32

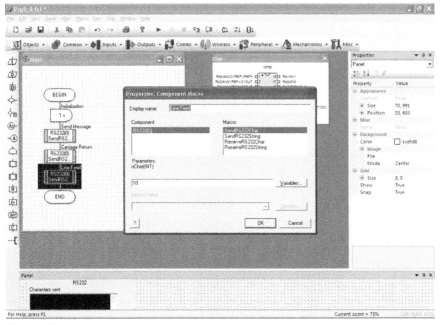

Figure 8.33

Next add a new Component icon after the Carriage Return icon; title this icon Line Feed. Open up the icon's Properties window (see Figure 8.33). Select the RS232(0) component, and then select the "SendRS232Char" in the macro window. In the Parameters text line enter the number 10 and hit OK.

Save and compile the program. Upload the firmware into your 16F88 PIC microcontroller and install it into your circuit on the prototyping breadboard.

Ready, Set and Go

Next connect your PC to the circuit through the DB9 (RS232-Serial) port. Start up your PC telecommunication program, HyperTerminal or PuTTY. Power up your program. If the COM port and communication parameters have been set up properly, upon powering up, the text messages shown in Figure 8.34 (for HyperTerminal) and Figure 8.35 (for PuTTY) will be typed on your communication screen.

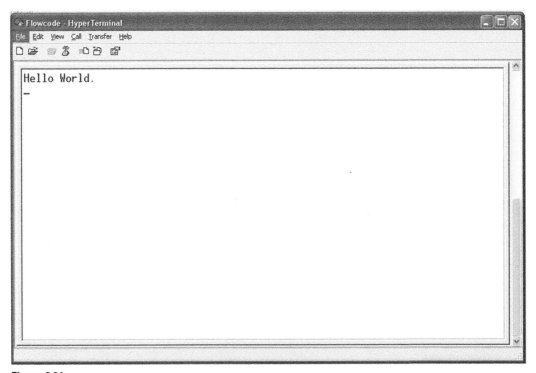

Figure 8.34

Figure 8.35

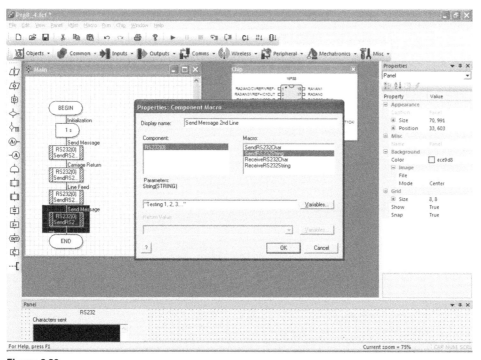

Figure 8.36

Hitting the reset button will print another text line with "Hello World." You can do that, or we can add a second line to our program. Now let's do this quick and easy. Select and highlight the Send Message icon. Next hit the two buttons "Ctrl" and "c" to copy the icon. Next select and highlight the Line Feed icon. When you do so, a small arrow will appear under the icon. This arrow shows where the copied icon will appear. This is what we want to do and where we want to place the icon, so hit the keys "Ctrl" and "v" to paste the icon.

Now you have two "Send Message" icons. Open up the Properties window of the icon you just pasted inside your Flowchart. Change the Display Name to "Send Message 2nd Line" and the text message to "Testing 1,2,3,..." (see Figure 8.36).

Save the program as 8.5; compile and upload the new firmware into your 16F88. Run the program. When you run the program it will type two lines (see Figure 8.37).

Figure 8.37

Establishing Two-Way Communication or Controlling the PIC from the PC

Now that we can send data to the PC, let's reverse the situation and send data from the PC to the PIC. Our first question is "What do we do with this data from the PC?" We could mirror the text data we receive on an LCD display but that is kind of redundant. It may have some application, and if that appears to be a worthwhile project for you, you have the information to bring that project to fruition. Rather I will have the PIC read all the data from the PC and when the PC sends the right code (character) I will have it turn on the LED for one second.

ECHO

We need to have our telecom programs echo the characters we type on the keyboard onto their communication screen. If we do not, the characters will be sent via serial communication, but we will not see them. So the next step is to set up the ECHO.

For HyperTerminal

Start HyperTerminal, go to the File menu and select Properties. This opens up the Properties window; in the "Settings" tab, click on the ASCII set-up button. This opens the ASCII set-up window. Under the ASCII Sending, set the "Echo typed characters locally" option and "Send line ends with line feeds." Hit OK to save settings, then hit OK again to return to the main screen (see Figure 8.38).

For PuTTY

Start PuTTY, set up COM3 and 1200 baud as you did before. Next click on the Terminal listing in the left hand side window. Set the "Local echo" force on option and "Local line editing" force on (see Figure 8.39).

Before opening the window, click on the Session Icon in the left hand window. This will bring you back to the original screen. Save the setting under a unique file name, I choose Flowcode. The next time you start PuTTY, you can simply load the settings instead of entering them. Once saved, hit Open to start your session in PuTTY.

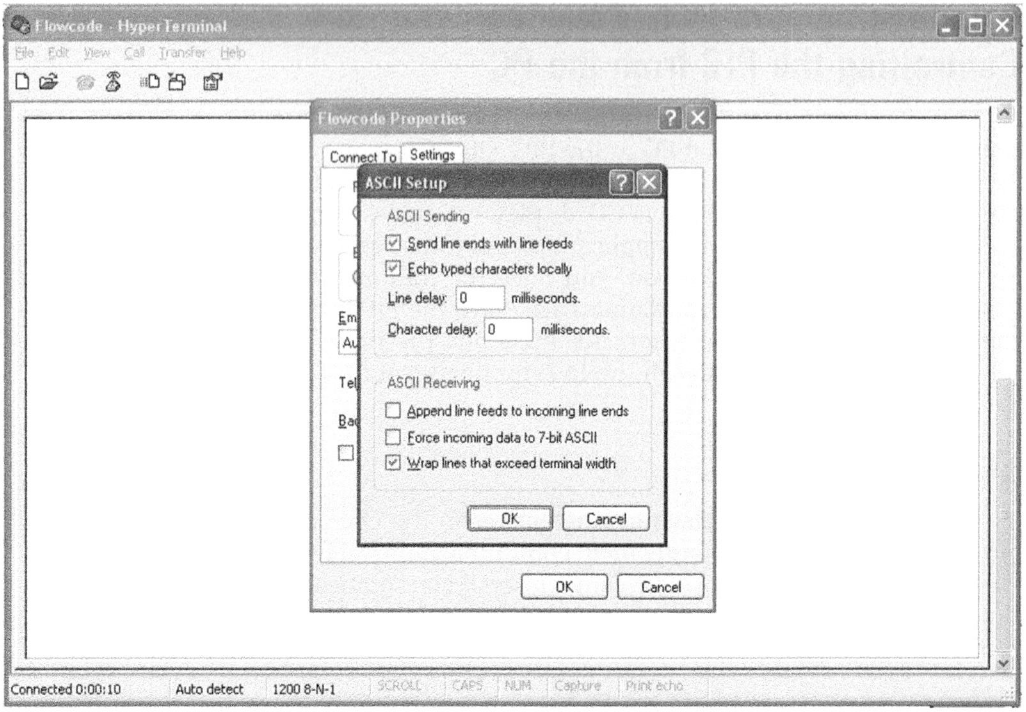

Figure 8.38

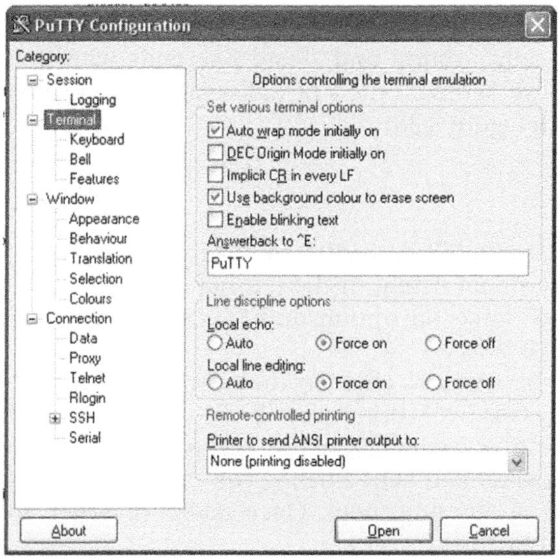

Figure 8.39

The Flowchart

Save your existing program as 8.6 before we begin modifying the program. Under the "Send Message 2nd Line" icon add a connection point. Under this add a Call Component macro. Title this icon "Receive," open its Properties window. Select the RS232 (0) component, then select the "ReceiveRS232Char" macro. Under the parameter "nTimeout(INT)" enter 1000. Under the "ReturnValue(INT)" click on the Variables button and create a byte variable named "num." Place this variable in the "Return Value(INT)" text box and click OK (see Figure 8.40).

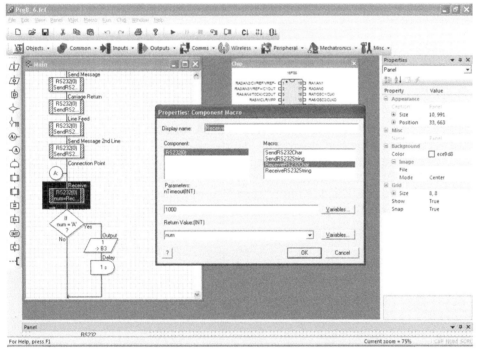

Figure 8.40

Next place a Decision icon. Open its Properties window. Make its Display name "Decision" and type in "num = 'A'" in the If text box, then click OK (see Figure 8.41).

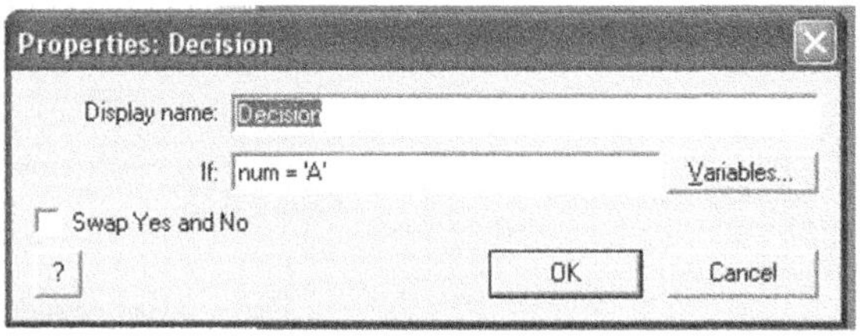

In the Yes branch of the decision, place an Output icon and open its Properties window. In the Display name

Figure 8.41

type "Output." Set value to "1." Select Port B, Single Bit and select 3 as shown in Figure 8.42. The bit 3 on Port B is where we connected our LED and resistor.

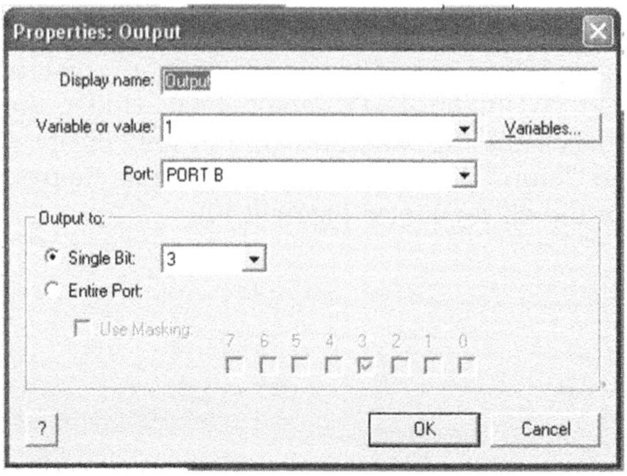

Figure 8.42

Remaining in the Yes branch, under the Output icon add a one second delay icon. Underneath the Yes branch, after it has rejoined the No branch and the main Flowchart line, place another Output icon. Open the Properties of this Output icon. In the Display name type "Output." Set the value to "0." Select Port B, Single Bit and select 3 as shown in Figure 8.43.

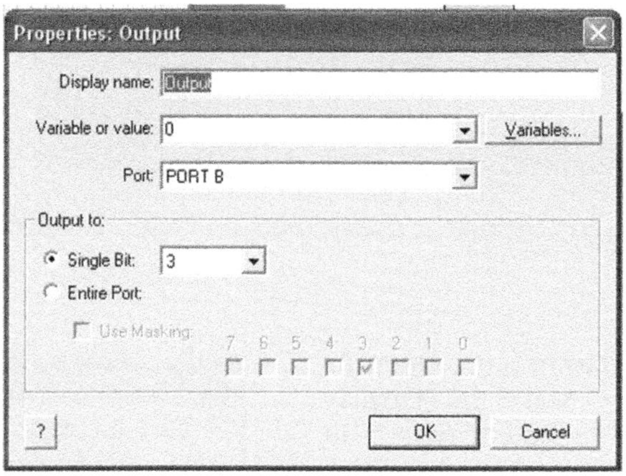

Figure 8.43

Save the program, compile, and upload firmware into your 16F88 microcontroller.

Controlling the PIC Microcontroller from the PC

When you run this program, it operates as it had before, printing two lines of text. The difference now is that when you type on the PC keyboard, the characters will show on the communication screen. If you type an upper case "A," the red LED connected to Port B will light for one second. Pressing any other key has no effect.

If your LED lights, you have established two-way communication between the PC and PIC; take a bow. While turning on an LED is not very significant in and of itself it does represent a universe of possibilities. For instance, you could nest character-based decision icons inside one another to create a character entry or word entry code, so that the user needs to enter the right sequence of letters in a specified time period to activate the LED. The LED in this scenario could represent the base of an NPN Darlington transistor that activates a solenoid-operated release on an electronic lock.

There is no limit to what the lighted LED could represent. It could represent anything, activating an external circuit, or executing another Flowchart program. This program is a like a tool. It's here for you if and when you decide to use it and how to use it.

UART Hardware

We established our PC link using the software control only because that's where the previous Flowchart program left off. This program can be modified to work just as well using the UART hardware built inside the PIC chip. In many cases the UART might be the better choice. In the next chapter we will begin our work with analog-to-digital converters, and we'll be using the UART again.

Expert Information and Notes for LCD Communication

Once in normal operation mode, a 5 millisecond delay should be kept after sending commands to the LCD, especially Clear screen, if not, following data/commands may be neglected. At higher baud rates (especially in inverted mode 9600 and above) a character pacing of 100 μs is required. This is the delay between consecutive characters, when multiple characters are sent in one serial command to LCD.

ANALOG-TO-DIGITAL CONVERTERS

An analog-to-digital converter measures a real world analog voltage (or current) and converts it to a proportional digital number that is equivalent in magnitude to that analog voltage. That digital number is a representative value that the PIC microcontroller (or computer) can read and use. Analog-to-digital converters may be abbreviated as: A/D, ADC, or A to D.

What I described in the preceding paragraph is the foundation of the digital recording industry, whether you talking about taking pictures in a digital camera, recording video in a DVR, or digital sound in a compact CD. These signals, as well as a host of other signals, are all digitized from analog sources using A/D converters.

For such important technologies we ought to define our quantities to be sure we understand "What is a real world analog voltage?". An analog voltage can be just about any AC or DC voltage you run across in the real world, for instance, voltage from a transistor battery, the output from a wall transformer or household voltage. These could also be electrical signals, like the electrical audio signal from a microphone, light intensity from a CCD array in a camera, or the electrical signal from any suitable sensor or transducer. What qualifies a voltage to be analog is that an analog voltage can vary in magnitude at a continuous rate.

The second part of our definition is Proportional Digital Number. Digital values can only vary in discrete values and increments, and can approximate the analog voltage. Figure 9.1 will help clear up any confusion this explanation created.

The line above the columns represents the analog signal (voltage), the columns are the equivalent digital measurement from an A/D converter. In this drawing the A/D only has 10 increments between 0 and +5 volts. This only provides a crude

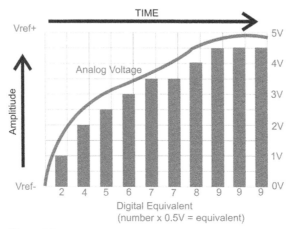

Figure 9.1

approximation of our analog signal. Our 16F88 PIC microcontroller has a built-in A/D converter. We will use the 16F88 converter in our first program. Our program reads the A/D with a resolution of 256 increments between 0 and 5 volts. This 256 increment gives a much better approximation of the analog signal than that shown in the illustration. We are not done with this illustration yet. Notice on the left side we have two quantities listed: −Vref and +Vref. These are values that are set up with your A/D converter. In this example the quantity −Vref is set at 0 volts and the quantity +Vref is set at +5 volts. So our A/D converter can read any voltage between these two quantities with a resolution of 8 bits (256 increments).

However, we can set −Vref and +Vref to other values. Suppose for whatever reason we are only interested in the voltage range between +2 and +3 volts. By setting −Vref to +2 volts and +Vref to +3 volts, the A/D converter can read the voltages between these two quantities with the same resolution of 8 bits. So while the range decreased from 5 volts to 1 volt, the resolution of the reading increased five-fold. We can calculate this with a little math. In our first example we have 5 volts/ 256 = 0.01953 volts per increment. In our second example we have 1 volt/256 = 0.00390 volts per increment. To see how much greater the resolution is, we divide 0.01953/0.00390 = 5. For restrictions on the use of Vref, you need to read the data sheets on that particular IC. One common restriction is that −Vref is a lower voltage than +Vref.

A/D Resolution

While we will work with an 8-bit converter, there are plenty of higher resolution converters available. The number of bits indicates the converter resolution. See the following table:

Bits	Resolution
4 bits	16
8 bits	256
10 bits	1024
16 bits	65,536
32 bits	4,294,967,296

The A/D converter in the 16F88 is actually 10 bits. We are reducing our reading of the A/D converter to 8 bits. By doing so, we are only looking at the 8 most significant bits on the conversion. Why? Because reducing the reading of the converter into one byte simplifies our programming. If one has a need to go to 1024 bit resolution it's still available using the 16F88.

Sampling Time

Time is shown in Figure 9.1 and is an important consideration. How many conversions or digital equivalent samples per second are taken from our analog signal? We can vary our sampling times to a point, in the Properties window of the A/D converter. For our programs here I would estimate we are sampling at around 10−25 samples per second, which is plenty fast for our applications. Just know that when sampling high fidelity sound or video signals, one would need to sample at much higher rates. For audio applications, think 40,000 samples per second and greater.

The 16F88 has one A/D converter but it may be accessed on multiple channels. What does that mean? It means we can sample electrical signals on more than one I/O pin on the 16F88. However, with each channel we employ our maximum speed-sampling rate is reduced in proportion to the number of channels.

First Program

We are going to incorporate what we learned in previous chapters in this chapter. We will take the output reading from our A/D converter and print it on an LCD screen. Now you have two choices for the LCD screen. You can use the parallel LCD interface from Chapter 7, or the serial LCD interface from Chapter 8. For simplicity's sake, the examples I am presenting in this chapter will use the serial LCD and interface. It is a simpler and cleaner (read fewer wires) schematic.

Vref

Our Vref quantities for the A/D converter are 0 volts or ground for −Vref and +5 volts for +Vref. The schematic for our program is shown in Figure 9.2.

We have introduced a new component in this schematic labeled V1. V1 is a potentiometer (typically called a pot); it is

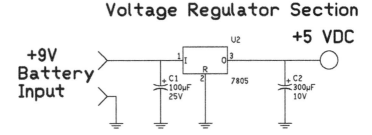

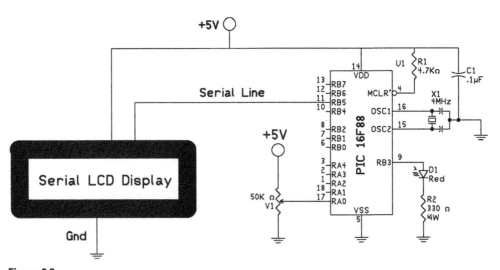

Figure 9.2

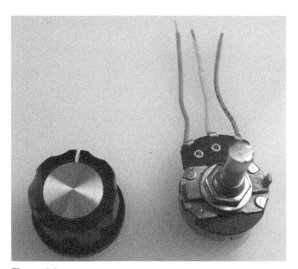

Figure 9.3

shown in Figure 9.3. A pot is a three-terminal variable resistor. By rotating the center shaft, the resistance of the center tap changes in proportion to the amount of rotation of the shaft.

There are two types of pot, linear resistance and logarithmic resistance. You want the linear style pot. This type of pot changes resistance in an even, linear fashion, as you would expect. The other types of pot are typically used in audio volume controls where the log change in resistance more closely changes the volume of an audio signal as your ears expect. The wiper (center tap) of the pot is connected to pin 17, RA0, which is also the AN0, analog input for the A/D converter. The resistance on my pot is 50 K, or 50,000 ohms. I chose this pot because it was handy. You could substitute other pot values such as 5 K, 10 K and 25 K. They will work equally well.

Potentiometer Function

The pot in the circuit behaves like two variable resistors in series connected to Vcc (+5 volts) and ground. The wiper is the center point between the two variable resistors. The two adjustable resistors form a voltage divider with a voltage appearing on the center point (wiper). As the shaft on the pot is turned, the resistors change value and the voltage that appears on the wiper of the pot changes in proportion. The voltage on the wiper can be varied between Vcc (+5 volts) and ground (0 volts). This program introduces a new component into the Flowcode chart called the A/D converter. Go to the Common menu and select A/D (see Figure 9.4).

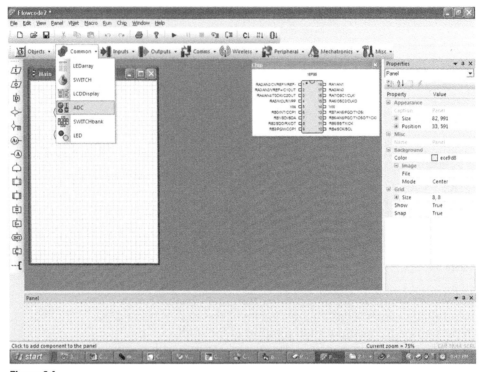

Figure 9.4

This places the A/D converter on your panel as a knob. Next add an RS232 component. Figure 9.5 shows our main program.

In the first Call Component macro, we select the ADC(0) as our component. Macro is ReadAsByte and create a one byte variable "num" that is placed in the Return Value[BYTE] text box (see Figure 9.6).

The value read from the analog-to-digital convertor will be stored in the num variable. Next we add a String Manipulation icon under the Call Component icon. This is to convert the num variable from a number to a string so it can be displayed on the LCD. Open up its Properties window as shown in Figure 9.7.

Create the variable "num_str" which is a 3 byte string variable (see Figure 9.8) and put the following into the String functions box:

num_str = ToString$(num)

Before we can place the next icon in the main Flowchart, create a new macro Flowchart named Clear_LCD_Screen. Go to Macro → New. This Flowchart is shown in Figure 9.9. This Flowchart uses four Call Component icons.

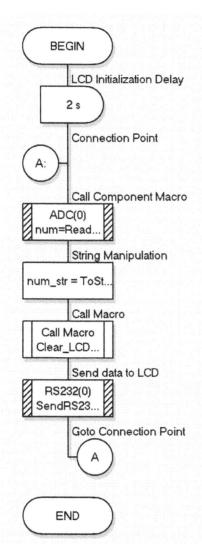

Figure 9.5

Figure 9.6

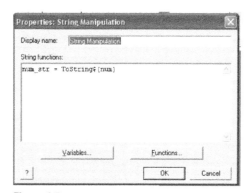

Figure 9.7

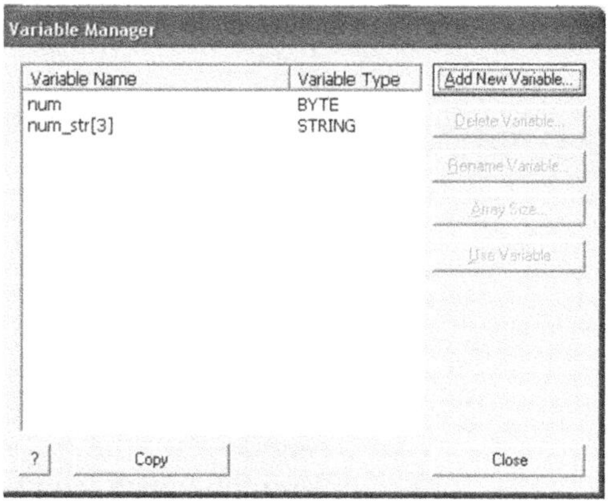

Figure 9.8

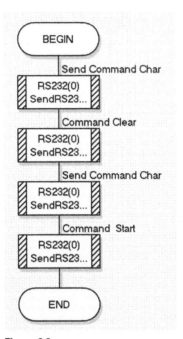

Figure 9.9

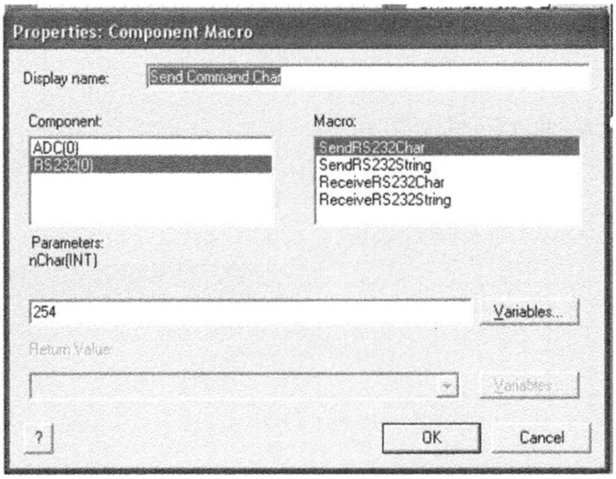

Figure 9.10

All icons call the RS232(0) component and use the SendRS232Char macro. The first icon sends byte 254 (see Figure 9.10) the second icon sends byte 0, the third icon byte 254 and the fourth icon byte 2.

Now let's set the properties of our RS232 component. Highlight the component in the Flow Code panel and select the External Properties to open up its Properties window (see Figure 9.11).

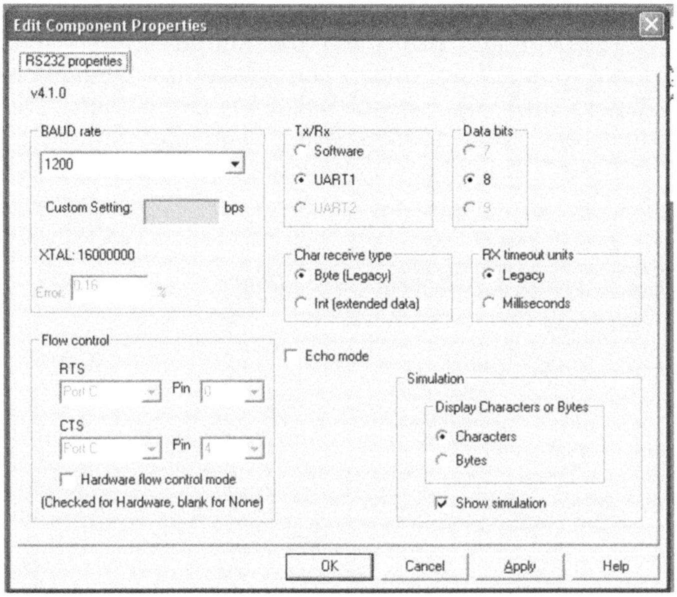

Figure 9.11

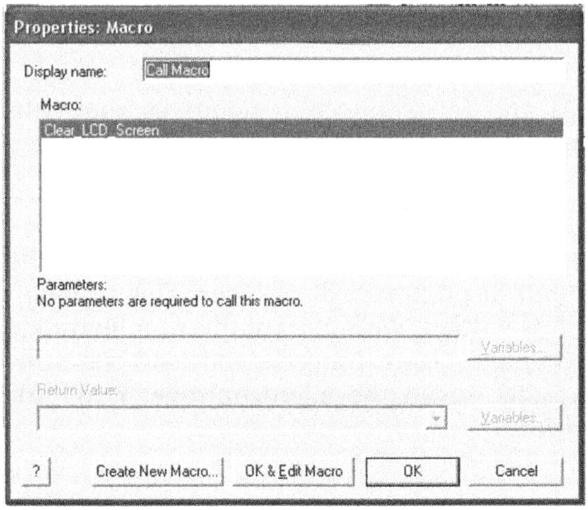

Figure 9.12

The circuit is designed to use the UART, so select the UART1 in the Tx/Rx option box. Select 1200 baud. Make sure you have the right crystal speed selected. We are ready to move forward with our main program (see Figure 9.5). Now we place the Call Macro icon under the String Manipulation icon and use it to call the Clear_LCD_Screen macro. The Properties window is shown in Figure 9.12.

The next Call Component icon in the main program calls the RS232(0) component and prints out the number "num" that was converted to a string value. The Properties window is shown in Figure 9.13. We use the RS232(0) component. The macro in this case is "SendRS232String" and the variable to send is "num_str." Title the Component icon "Send data to LCD" and hit OK.

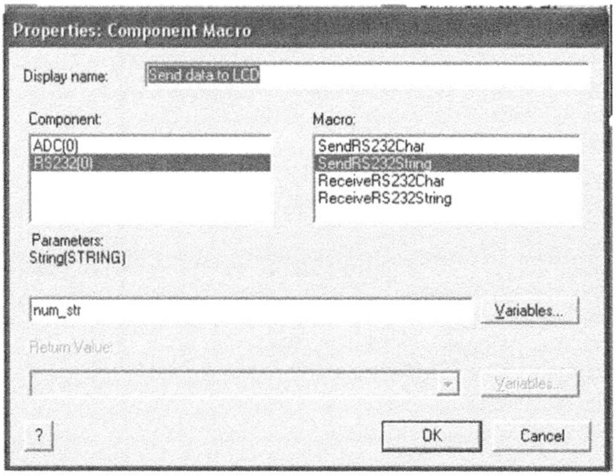

Figure 9.13

Place our two (A) and (A:) connection points as shown in Figure 9.5 and you're ready to compile the program.

Program Function

The program reads the voltage present on the wiper of the pot and converts it into a proportional number. This program will not simulate as it is written. We'll change this to a Flowcode chart that may be simulated later. Compile and load the firmware into your PIC16F88. Insert the microcontroller into your circuit and apply power.

Running the Program

When you apply power to the circuit a number will appear on the LCD screen. Rotating the pot's shaft will change the number displayed on the LCD up or down depending on which way you rotate the shaft. That number represents the voltage on the wiper. To determine the voltage that the number represents, go back to the beginning of this chapter. I stated that we could calculate this voltage with a little math. In our example we have 5 volts/256 = 0.01953 volts per increment. We multiply the number displayed on the LCD by 0.01953 volts to obtain the voltage measurement. So if the number displayed is 120, we take $120 \times 0.01953 \text{ V} = 2.34$ volts.

Getting Practical

So at this point you might be saying to yourself, that's nice, but what can I do with this technology? In a word, plenty! The A/D converter can read the analog voltage output of electrical equipment. That signal could represent just about anything an engineer would want to know about or monitor. For us, there are a variety of passive resistive sensors that change their resistance in response to energy. We can switch out our potentiometer and replace it with any of these passive resistive sensors. The common sensors available respond to:
• Light
• Heat (temperature)
• Magnetic fields
• Bend
• Stretch
• Gas (toxic)
• Alcohol
• Water
• Humidity.

In addition to sensing these energies, we can have the microcontroller respond to the energy it detects or to the changes in energy. A simple response that we will use to show detection or threshold is lighting an LED. So before we start hooking up sensors, let's add a few commands to give a visual indication. Our circuit already has an LED connected to RB3 for an indicator, so all we need to do is update our program. Resave the program as 9_2.fcf before updating (see Figure 9.14).

We want to read the value of the "num" variable. So add a Decision icon below the Component macro "Send data to LCD." In the Properties window of the Decision icon add num > 127 in the If text box as shown in Figure 9.15. I chose the number 127 because it is the approximate midpoint (2.5 volts) of the voltage range of our A/D reading range from 0 volts to 5 volts. This threshold point number can be changed to suit the transducer and range.

In the Yes branch of the Decision icon, we add an Output icon. We open its Properties window and the set its properties as shown in Figure 9.16. Using this output in the Yes branch

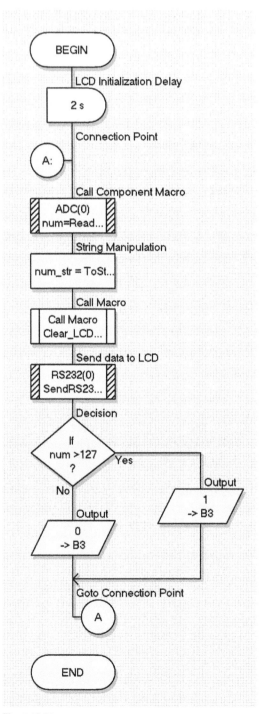

Figure 9.14

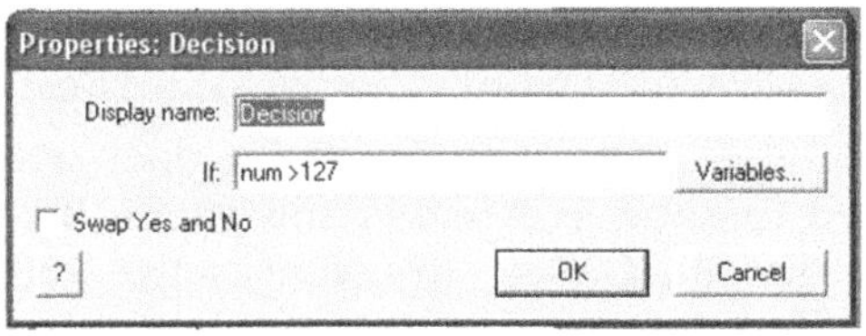

Figure 9.15

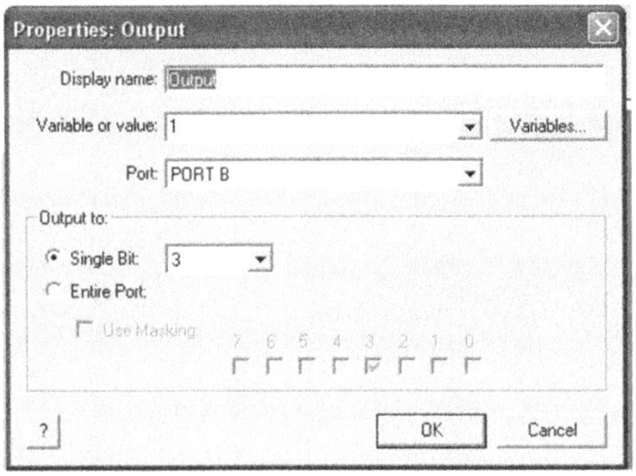

Figure 9.16

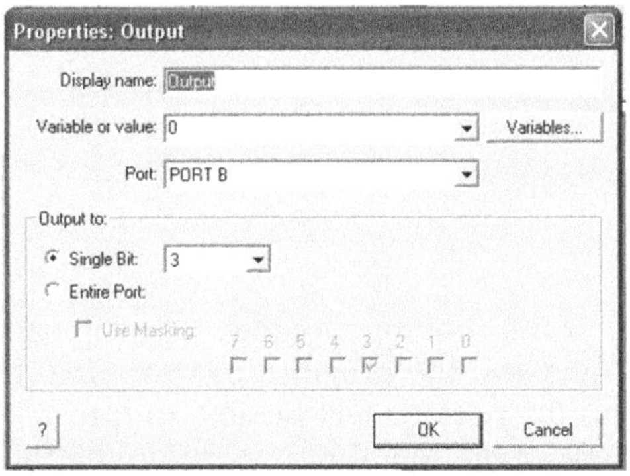

Figure 9.17

creates the function so that if the number "num" from the A/D is above 127 the LED will turn on.

In the No branch of the Decision icon, we add an Output icon. We open its Properties window and then set its properties as shown in Figure 9.17. Using this output in the No branch creates the function so that if the number "num" from the A/D is below or equal to 127 the LED will turn off.

Save, and compile your program to a Hex file.

Special Considerations when Using RB3 for Digital I/O

Low voltage programming is the default mode for the 16F88 chip. What low voltage programming is at this point does not concern us. What we need to know is whether this mode disables the RB3 for use as a digital I/O. Since we want to use RB3 for digital I/O we MUST disable the low voltage programming in the chip. To do so, we select the disable low voltage programming in the EPICs (programmer) setting before we upload our firmware into the 16F88 (see Figure 9.18).

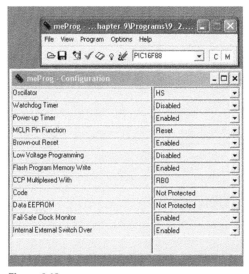

Figure 9.18

Use the setting shown in the meProg window; then upload the program into the PIC and run the program in your circuit. When you turn the shaft on the pot you will be able to turn the LED on and off depending upon the rotary position of the potentiometer shaft. This threshold ought to be around the midpoint rotary position of the shaft. The A/D number displayed on your LCD will tell you exactly when the LED will turn on and off. It will remain off for all numbers under 127 including the number 127. For all numbers above, from 128 on upwards, the LED will turn on (see Figure 9.19).

Before we move on to adding sensors to our project, let's look at the program again. I use connection points A and A: to jump around and create loops inside our program. This goes back to my GWBasic days, when I used GOTO statements to jump around inside a Basic program. The Flowcode connection points work in a similar way. My point in bringing this up is that you

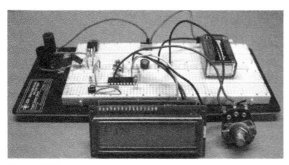

Figure 9.19

can use the Loop-While icons to accomplish much of the same thing.

Loop-While

Resave your current program as 9_3.fcf. (see Figure 9.20).

In this revision, add the Loop-While icon under the 2 second LCD Initialization icon. Next, remove the two connection points, A: and A icons. Now copy and paste all the remaining icons between the Loop icon and End icon and paste them inside the Loop-While as shown. The Loop-While icon loops the program until the specified condition becomes true. You can enter the conditions which will fulfill the loop. In our case, we are using the default setting. In this setting the test condition is to something that is always true and will make the loop repeat forever, e.g., While 1. Compile and test the program. This Loop-While version will function the same as will the A Connection points program.

Loops and Jumps

While both programs function in an identical manner in this particular program, can you think which case you would use Connection points in, in contrast to Loop-While? Loop-While provides an opportunity to set a loop condition. The Connection points in contrast are a hard loop – when encountered they always take action. Also with Connection points, we have the opportunity to set multiple A points to jump to a single A: point in the program.

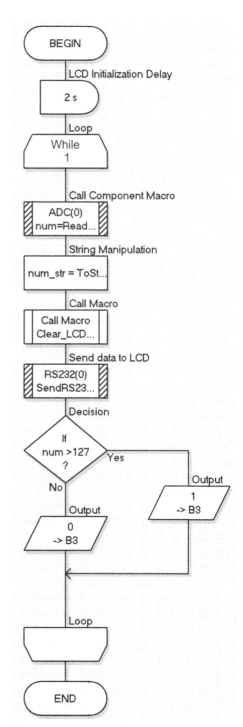

Figure 9.20

Simulate or Not Simulate

As the program is currently written it will not simulate properly in Flowcode. Can we modify the program so that we can see the actions simulate? Let's give it a try. The first step is to resave the program as 9_4.fcf. Next remove the Call for the Clear_LCD_Screen. That entire macro can be removed. Under the "Send data to LCD" Component Macro, add another Component Macro. Set Properties values to: Title "Add Blank Char Space," Component RS232(0), Macro "SendCharString" and " " in the Variables text box as shown in Figure 9.21.

We only need one more component to add to our program and that is an LED. Add a single LED component and open up its Connection Properties as shown in Figure 9.22.

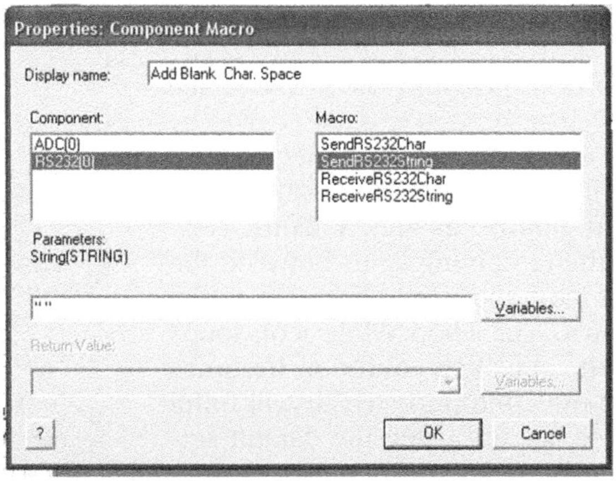

Figure 9.21

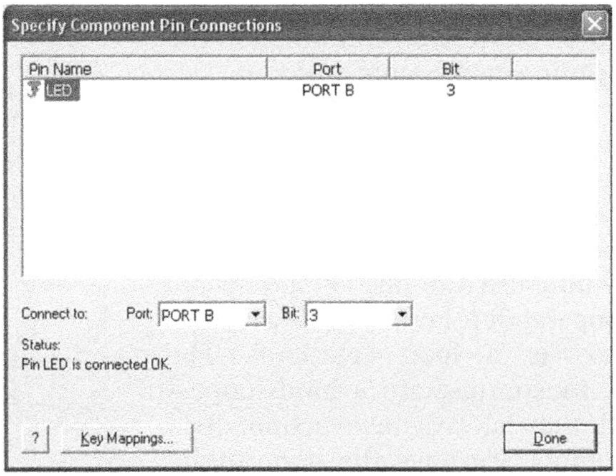

Figure 9.22

We connect the LED to Port B on bit 3 then click on Done. At this point you can run the simulation. The current A/D value will print in the RS232 Character Sent window. You can use your mouse to turn the knob on the potentiometer connected to the A/D, and see the numbers change in the RS232 Character Sent window. When the number reaches 128, the LED on the simulation panel will light up. Go below 128 and the LED will turn off. The simulation mimics our real world circuit. Notice that we didn't have to set up the knob and potentiometer to the A/D converter. This is the default setting.

Sensors

With this experience under our belts we can implement real world sensors to out circuit. We are going to connect four different passive resistive sensors: Cadmium sulfide (CdS) photo resistor, Thermistor (heat sensitive resistor), Bi-flex bend sensor and Stretch sensor (see Figure 9.23).

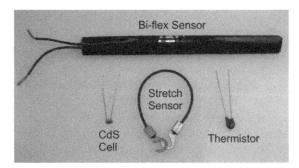

Figure 9.23

Notice the sensors are all two-terminal devices, in comparison with the potentiometer, which is a three-terminal device. In order to read the change in our sensor's resistance, we will place our sensor between two resistors in series, as shown in Figure 9.24. The resistors R3 and R4 can be changed to compensate for the nominal resistance value of the sensor and the change in the sensor's resistance response.

Nominal Sensor Resistances

The nominal resistances for the sensors are as follows:
* Bi-Flex bend sensor 7 K ohms
* Thermistor 11 K ohms (74° F)

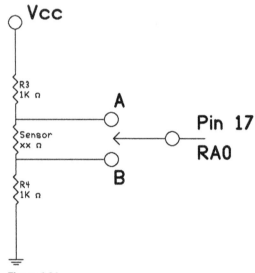

Figure 9.24

- CdS cell 100 K ohms (Dark); 2 K ohms (in light)
- Stretch sensor 2.5 K ohms.

The Bi-Flex bend sensor decreases in resistance when the sensor is flexed or bent. The sensor is also pressure sensitive. So applying pressure will also reduce its resistance. We can place the Bi-Flex sensor in the sensor position as shown in Figure 9.24. We have two connection points: A and B. We select only one connection point to connect to pin 17 of our micro-controller. We select the connection point depending upon how we want to read the voltage output from the sensor.

Ohm's Law

To calculate the voltage response of our sensor and its reading from the A/D converter we need to use Ohm's law. These equations use three symbols. E represents volts, I represents current and R represents resistance. Ohm's law consists of three basic equations:

$$E = IR, \ I = E/R \ \text{and} \ R = E/I$$

And that's all we need to do this work. Let's start using Ohm's law by continuing with the Bi-Flex sensor. Its nominal resistance is 7000 ohms, R3 is 1000 ohms and R4 is 1000 ohms.

Adding our resistances together we have Rt = 9000 ohms. We can use that value to calculate the current. We know our voltage is 5 volts. Plugging the numbers in, we have:

$$I = 5 \ V/9000 \text{ ohms}$$

so I = 0.00055 or approximately 0.55 mA (milliamps)

We can use this current to determine the voltage the A/D will see at points A and B. At point B the current is only flowing through the 1000 ohms of R4 to ground. We know the current, so using Ohm's law E = IR we can calculate the voltage: E = 0.00055 × 1000 = 0.55 volts. So at point B the A/D would read a little bit above one half of a volt.

We can use the same equation to calculate the voltage at point A. The resistance at point A is 7000 (sensor) plus 1000 (R4) = 8000 ohms. E = 0.00055 × 8000 = 4.44 volts. So at point A the A/D would read about four and one half volts. We can use these voltages to approximate what numbers the A/D would output from reading the voltages at points A and B. We know our reference voltages for the A/D converter are 0 volts (ground) and +5 volts. From this we have an 8-bit resolution or 256 steps from 0 volts to 5 volts. So each number increment has a voltage value of (5 volts/256 steps) 0.0195 volts.

- At point A (a voltage of 4.44 volts (4.44 V/.0195)) the A/D will output a number of 227.
- At point B (a voltage of 0.55 volts (0.55/0.0195)) the A/D will output a number of 28.

What happens when we bend the Bi-Flex sensor? If we bend the sensor by 90°, its resistance drops to 4000 ohms. This changes the whole resistive system dynamically. What will the A/D read at points A and B? To determine this we need to recalculate all our values. Rt, which was 9000 ohms, has now become 6000 ohms. So the current through the resistors and sensor that was 0.55 mA is now 0.83 mA. Therefore the voltage at point B is now 0.83 volts and that at point A is 4.15 volts.

- At point A (a voltage of 4.15 volts (4.15 V/0.0195)) the A/D will output a number of 212.
- At point B (a voltage of 0.83 volts (0.83/.0195)) the A/D will output a number of 42.

Fortunately, you do not need to calculate the changes. You can just connect your sensor and resistors into the circuit and read the results on the LCD. The math gives you an idea of what to expect.

Real World Numbers

Let's do a quick comparison with the calculated results to what I measured in the real world.

At point A			
Nominal Resistance	:	Calculated 227	Real World 234
90° bend	:	Calculated 212	Real World 210

At point B			
Nominal Resistance	:	Calculated 28	Real World 23
90° bend	:	Calculated 42	Real World 32

Let's analyze the calculated results a little further. With a 90° bend in the sensor we are seeing a difference of 15 at point A (227–212) and a change of 14 at point B. What could we do to see a bigger numerical difference as the sensor is bent?

Setting + Vref and − Vref

Modifying the resistor would help a little, but not a lot. The key is in changing the reference voltages (Vref+ and Vref−) for the A/D converter. Using two pins on the 16F88 for Vref is software selectable. One needs to modify the ADCON1 register. When the ADCON1 is properly configured the Vref+ is attached to RA3 (pin #2) and Vref− is attached to RA2 (pin #1). You would have to choose which point (A or B) you wanted to select proper reference voltages. Let's assume we choose point A. Your voltage range from the sensor is 4.15 volts to 4.44 volts. If you made +Vref = 4.5 volts, a little higher than 4.44 volts, and −Vref = 4.1 volts, which is a little less than 4.15 volts, you can be comfortable that your sensor voltage will not go off scale and out of range. Setting this voltage range, the resolution from you A/D converter is $((4.5\,V - 4.1\,V)/256)) = 0.00156\,V$ per increment, starting at 4.1 volts. Therefore, the A/D output at the sensor's nominal resistance is:

$$((4.15\,V - 4.1\,V)/0.00156) = 32.$$

The A/D output when the sensor is bent at 90° is:

$$((4.44\,V - 4.1\,V)/0.00156) = 217.$$

Now our A/D numerical output varies by 185 as the sensor is bent. A much higher resolution can be achieved using the Vref voltage than can be accomplished by resistors alone.

Enabling +Vref in Flowcode

Enabling +Vref may be accomplished in the External Properties of the A/D converter. Select the A/D known in the panel and open its External Properties window (see Figure 9.25).

Select VREF+ in the Vref Option box and click OK. Save, compile and reprogram the PIC microcontroller in your circuit. To adjust the Vref+ on Pin 2, we connect the wiper of a 50 K pot to the pin, as shown in Figure 9.26.

Rotating the shaft changes the voltage presented on the Vref+ pin. You can measure the voltage on Vref+ using a VOM meter set to volts. When I present +1 volts on the Vref+ pin I obtained the following numerical output from the A/D converter:

At point B (0.25 volts)		Standard	Using Vref + (+1 volt)
Nominal Resistance	:	23	30
90° bend	:	32	100

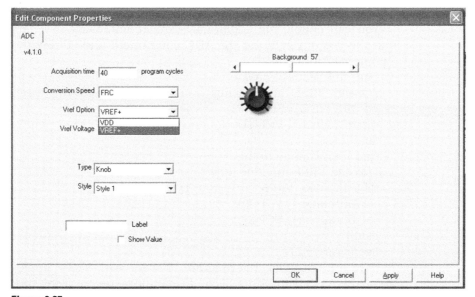

Figure 9.25

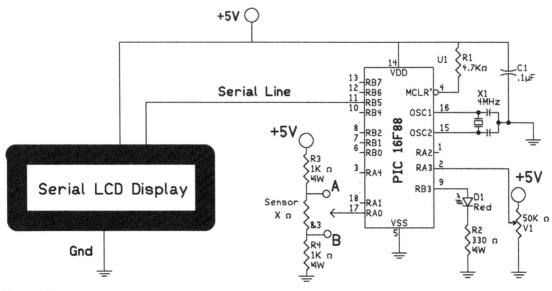

Figure 9.26

As you can see we have a substantial improvement in our reading range by implementing Vref+.

Enabling — Vref in Flowcode

You can enable the additional −Vref functionality by going into the custom code for the ADC component and editing the SampleADC function.

Find this section of code:

```
//assign VREF functionality
#if (MX_ADC_VREF_OPT = = 1)
set_bit(adcon1, VCFG1);
#endif
```

And then change to this

```
//assign VREF functionality
#if (MX_ADC_VREF_OPT = = 1)
set_bit(adcon1, VCFG0);
set_bit(adcon1, VCFG1);
#endif
```

See Figure 9.27.

Save and compile this program change. With Vref− implemented we need to add a second potentiometer to Vref− (pin #1) to our circuit (see Figure 9.28).

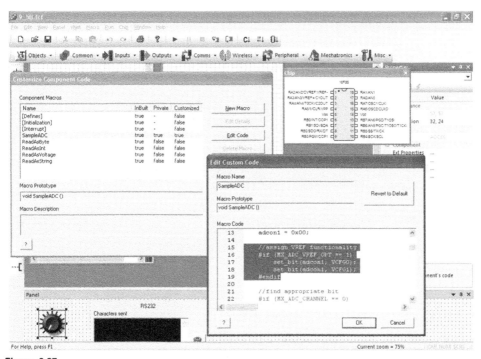

Figure 9.27

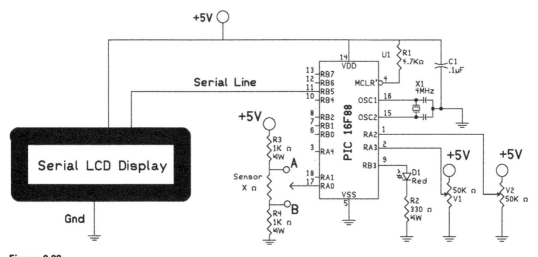

Figure 9.28

I set Vref− to 3.8 volts and Vref+ to 4.5 volts. Connecting the A/D input (RA0) to the A connection point and adjusting the Vrefs as specified I was able to vary the numerical output of the A/D converter by flexing the bend sensor between 200 (nominal) and 0.

10

POWER CONTROL FROM DC MOTOR TO AC APPLIANCES

There are different levels of DC power control from small motors up to and including main power from your home. We will start small, using DC hobby motors and ending with controlling full power appliances for your home.

Controlling DC Power

In Chapter 3 we began learning Flowcode programming by turning on and off LEDs in our program "Wink." The same voltage output that we used to turn on an LED can also be used to turn on a transistor. The transistor in turn may be used to control far more current than that which could be supplied by our microcontroller's I/O pin.

Using a Transistor as a Switch

A standard bipolar junction transistor (BJT) comes in two main flavors, NPN and PNP. The N in the transistor type indicates an N-type semiconductor and P is for a P-type semiconductor. So the NPN indicates the P-type semiconductor material is sandwiched between two N-type semiconductor materials. While transistors are commonly used for amplifying signals, we will be operating them at saturation where they behave more like switches.

In Figure 10.1 we are using a common NPN transistor as a switch. Our microcontroller will output +5 volts on its RB0 and RB1 I/O pins. The reason for the LED connected to RB0 will become apparent as we work through our examples. When we output the +5 volts on RB0 it will light the LED connected to RB0; the +5 voltage on RB1 turns on the transistor, which in turn will allow the LED to be lit. When you run our program, you will

PIC Projects for Non-Programmers. DOI: 10.1016/B978-1-85617-603-3.00010-5

Voltage Regulation

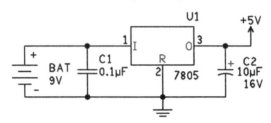

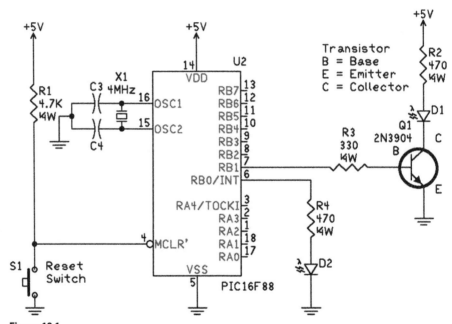

Figure 10.1

see that both LEDs light simultaneously. The advantage with using the transistor is that it can control more current than that which the microcontrollers I/O pin could supply. We will see this a little later when we connect a DC motor into our circuit.

The program is shown in Figure 10.2.

In Figure 10.3 we have an almost identical circuit. The single component we are changing is the transistor. It has gone from an NPN in Figure 10.1 to a PNP transistor in Figure 10.3.

So what does this mean in terms of function? In Figure 10.1 the NPN transistor is turned on when the I/O pin outputs 5 volts, and turned off when the I/O pin outputs 0 volts. So both LEDs in the circuit are lit simultaneously. In Figure 10.3, the PNP transistor is turned on when the I/O outputs 0 volts and

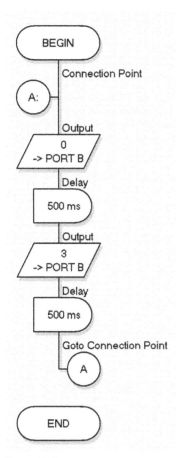

Figure 10.2

turned off when the I/O pin outputs 5 volts. The results when you run this circuit using the same program are that the LEDs will light alternately. When you are designing circuits you have the option of using either an NPN or a PNP transistor depending upon which signal you want to use to control the current. In Figure 10.4 we are switching out the transistor controlled LED and putting a DC hobby motor in its place.

In Figure 10.4 we have a reversed diode across the power connections to the motor. This is a snubber diode that dissipates the back EMF from the DC motor when it is turned off. Without the diode, the back EMF could burn out the transistor. Using the 2N3904 transistor and motor combination is only good for high efficiency DC motors. Most small hobby motors are grossly inefficient, requiring gobs (that's the technical term) of current to run. In Figure 10.5, we show a number of hobby motors and two gearbox motors. The two high efficiency motors are labeled.

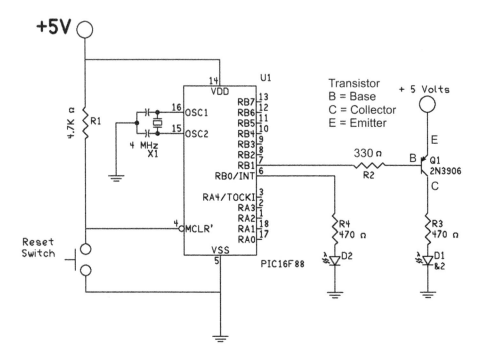

Figure 10.3

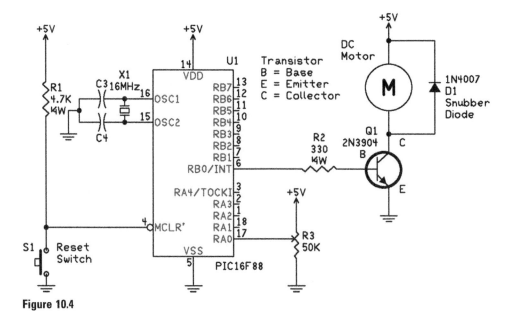

Figure 10.4

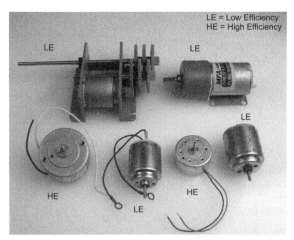

Figure 10.5

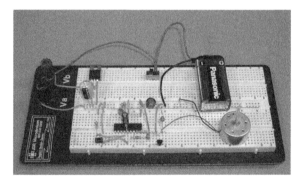

Figure 10.6

If you have these low efficiency hobby motors they will work, but not as well as a high efficiency motor like the ones used in solar energy applications or CD players.

Figure 10.6 shows my circuit. I substitute a subminiature red LED for the 1N4007. You can just barely see the LED flicker slightly when the power is cut from the motor. That's the back EMF. I placed a piece of scotch tape on the motor's shaft so I could see it rotate more easily.

H-Bridge

An H-bridge uses four transistors like switches and allows you to control the directional rotation of the DC motor. It's called an H-bridge because the transistors are arranged in a bridge-like pattern (see Figure 10.7).

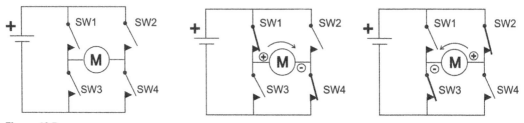

Figure 10.7

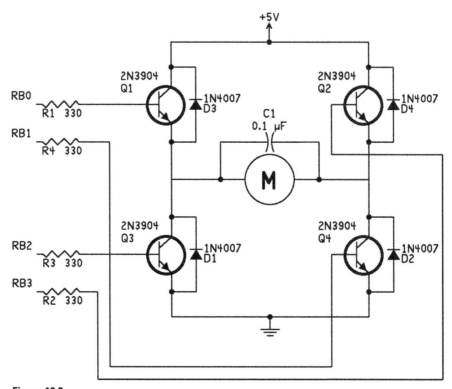

Figure 10.8

When switches SW1 and SW4 are closed, the motor shaft rotates in one direction. When switches SW2 and SW3 are closed, the motor shaft rotates in the opposite direction. When the switches are open, the motor has no power and its shaft does not rotate. We can make an H-bridge using our 2N3904 NPN transistors (see Figure 10.8). Now this is an inefficient H-bridge design (it is shown here for illustration purposes); it requires a high efficiency DC motor to function. In real world applications you can purchase ready to work H-bridges that work with higher efficiencies.

We need to modify our Flowcode program slightly. In this program we output +5 volts on pins RB0 and RB1 (Port B = 3) to rotate the motor's shaft in one direction for three seconds. Then we output +5 volts on pins RB2 and RB3 (Port B = 12) for three seconds to rotate the motor's shaft in the opposite direction. The cycle then repeats (see Figure 10.9).

The constructed circuit on the solderless breadboard is shown in Figure 10.10.

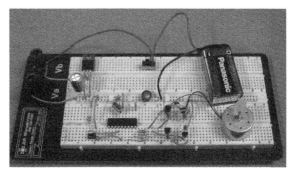

Figure 10.10

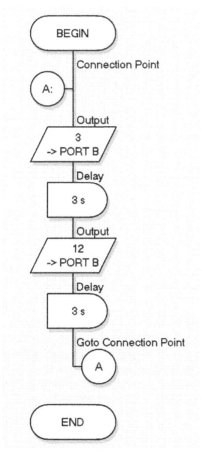

Figure 10.9

Another H-Bridge

The following schematic in Figure 10.11 shows another H-bridge configuration that uses complementary NPN and PNP

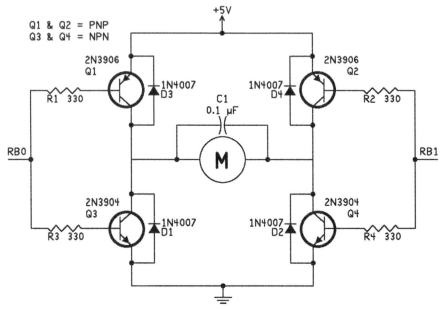

Figure 10.11

transistors. The advantage of this H-bridge is that a single I/O line controls direction, rather than using two I/O lines as in our previous example.

Motor Speed Control

We will close this section with creating a DC motor control. There is more than one method of controlling a motor's speed. Some people try to limit the current going to a motor with a variable resistor or a variable resistor connected to a power transistor. These methods work, but waste energy in the form of heat. The method we will use is called pulse width modulation (PWM). Pulse width modulation is easy to understand, if you look at Figure 10.12.

The lines illustrate the power being applied to our DC motor. In the first line power is applied continuously, or for 100% of the time. In the line underneath, power is applied 75% of the time using a square wave. Following that line is the next line that represents power being applied 50% of the time, followed by 25% of the time. These percentages are called duty cycles. If one said the duty cycle of the PWM signal is 50%, one would be saying that power is applied 50% of the time.

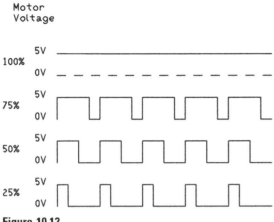

Figure 10.12

Time Period and Frequency

The frequency of the PWM is important. Our first program in this chapter could be considered a 50% duty cycle, with a 1 second time period. Obviously turning a motor on–off at 0.5 second intervals gives a very choppy performance. In general you want to use the lowest frequency that will provide an even flow of current to the DC motor. This will result in smooth motor performance and you can vary the speed of the motor. This frequency will vary from motor to motor, depending upon the particular specifications of the motor.

Fortunately for us, changing frequency and duty cycle is easy. Now there are a number of thoughts on the range of frequencies one ought to use. Some believe the frequency should be above the standard range of human hearing of 20 kHz. Others feel the 500 Hz to 1500 Hz range is fine. We will test both. To simplify our testing we are going back to our original single NPN transistor motor schematic and adding a potentiometer to our A/D converter. We will read the potentiometer with the A/D converter and use that to adjust the PWM signal duty cycle from 0% to 100%. Start a new Flowcode chart. Add the A/D control component as we did in the last chapter. Next add the PWM component available under the Mechatronics menu (see Figure 10.13).

Next add a Call Component icon, and open up its Properties window. Select the PWM(0) component. Then select the Enable macro. In the Parameters text field add the number 1 and hit OK (see Figure 10.14).

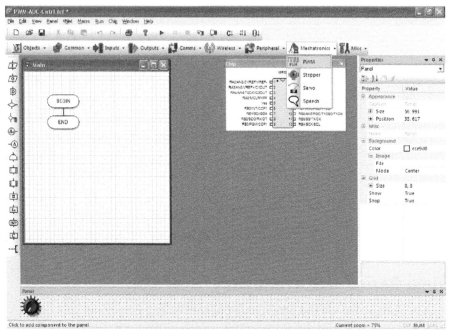

Figure 10.13

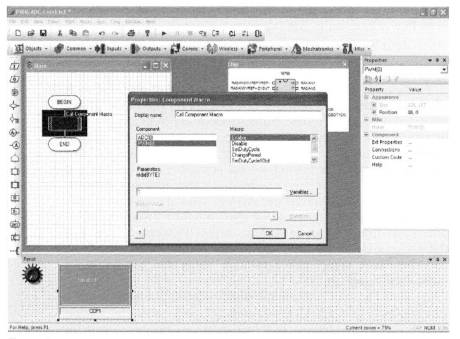

Figure 10.14

Next add While 1 loop (see Figure 10.15).

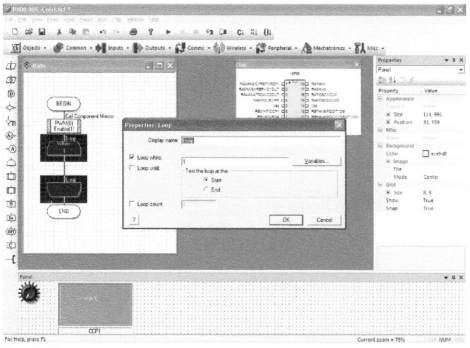

Figure 10.15

Inside the loop add a Call Component icon. Open its Properties window, and select the ADC(0) component. Select the ReadAsByte macro. In the Return Value (BYTE) text space, create a byte variable and call it x, and place it into this text space (see Figure 10.16) then click OK.

Add another Call Component icon. Open its Properties window. Select the PWM(0) component. Select the SetDutyCycle macro. In the nIdx(BYTE), nDuty(BYTE) parameters text box enter 1,x as shown in Figure 10.17, then click OK.

Next open up Project Options under the view menu. I have a 16 MHz crystal in my circuit, so I set the clock speed to 16 MHz (see Figure 10.18).

At this point we can run a simulation. Run the simulation and turn the knob to the A/D converter. As you do so, the duty cycle of the waveform in the green PWM space will change in proportion to the position of the A/D converter knob (see Figure 10.19).

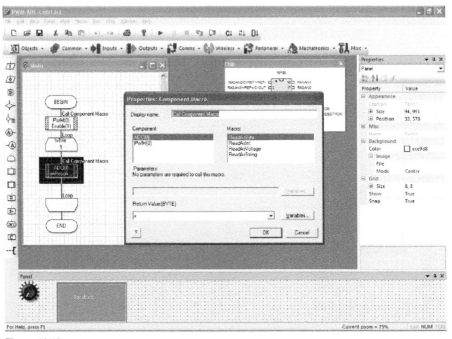

Figure 10.16

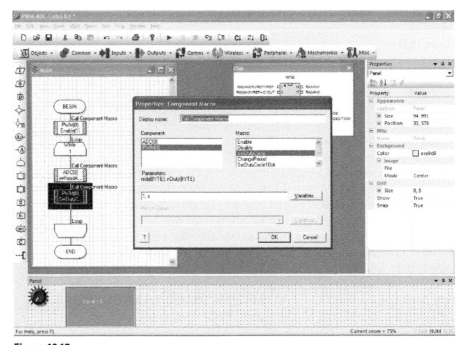

Figure 10.17

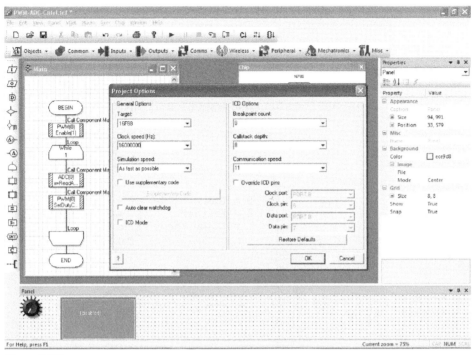

Figure 10.18

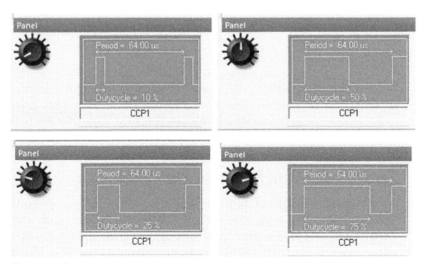

Figure 10.19

The frequency shown in Figure 10.19 has a period of 64 microseconds. Frequency = 1/(time period), so in this case it calculates to a frequency of 1/0.000064 or 15,625 Hz, or about 15 kHz. Compile the program to Hex, and upload the firmware into the 16F88 microcontroller. The PWM output pin for the 16F88 is pin 6. We will use the PWM output from pin 6 to control the speed of a DC motor. The schematic is shown in Figure 10.20.

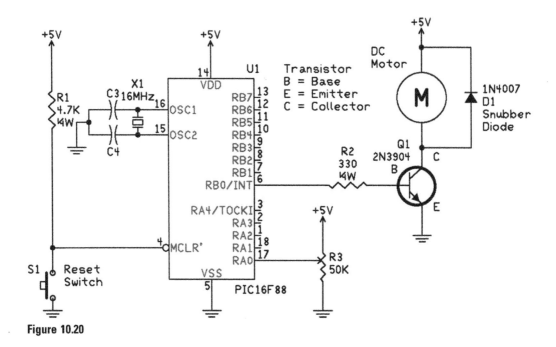

Figure 10.20

Run the program. By varying the resistance of the potentiometer, you can control the speed of the DC motor. This program is operating at about 15 kHz. In the next example we will adjust the frequency.

Setting Up Our Parameters

We used the default settings in our simulation; this default setting works OK with the A/D converter. However, with the PWM component it's time to go in and set up the frequency parameters. Select the PWM component icon in the panel and then select External Properties (see Figure 10.21). Choose the Clock source of clk/16. With a 16 MHz crystal this will provide a frequency of approximately 1000 Hz (1 kHz).

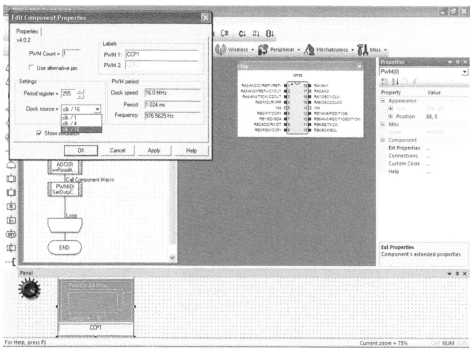

Figure 10.21

Run this program in your circuit. See if the frequency changes the DC motor's response in your circuit.

Visual Speed Control

In our examples we are using a visual indication of the speed of our DC motor. This is not always possible. Engineers have worked out a few methods of measuring and controlling the speed of a DC motor. While we will not implement any of these methods, I do want to mention them to you for further research. One method measures encoders. As the shaft rotates it breaks a light path to a sensor and the microcontroller can count the pulses to measure speed and/or rotation of the motor's shaft. Another technique uses Hall sensors and magnets in a similar manner by counting pulses. Other methods measure the current drawn from the motor to gauge its speed, and yet another technique measures the voltage from the back EMF from the motor.

Controlling Mains Power Supply

We can use the same technology to control main power supply for household electrical devices. Instead, on powering a DC motor, we control a relay switch. The switch of the relay can control household electrical devices (see Figure 10.22).

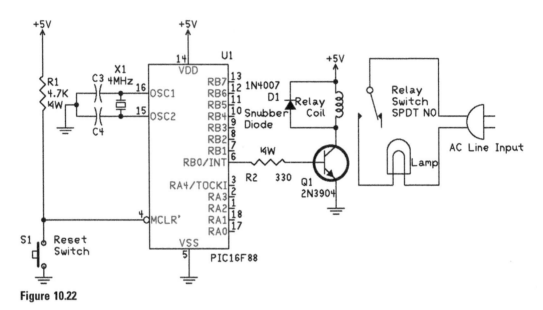

Figure 10.22

APPENDIX A
INSTALLING ECIO DRIVERS

If you're here I have to assume your Windows operating system did not find the driver for the ECIO hardware. You may be in Chapter 2 trying to program your ECIO28 or ECIO40 for the first time; here's what to do.

1. Visit the Matrix Multimedia website on the Internet.
2. Find the software downloads page (see Figure A.1).
3. Go to the "ECIO Programming Tool and USB Drivers" section on the page (see Figure A.1).
4. Download the USB ECIO drivers to your computer's hard drive (see Figure A.2).
5. There are three USB ECIO install driver programs. Ignore the install driver programs with the (amd64) and (ia64) suffixes (see Figure A.2). Run the ECIO install driver program (see Figure A.3).
6. Follow the screen directions; agree to the EULA software agreement (see Figure A.4).
7. If a Windows security screen pops up, allow the installation to continue (see Figure A.5).
8. The drivers are installing (see Figure A.6).
9. The finish screen (Figure A.7) shows that the drivers are installed. Go back to Chapter 2 and continue programming your ECIO module.

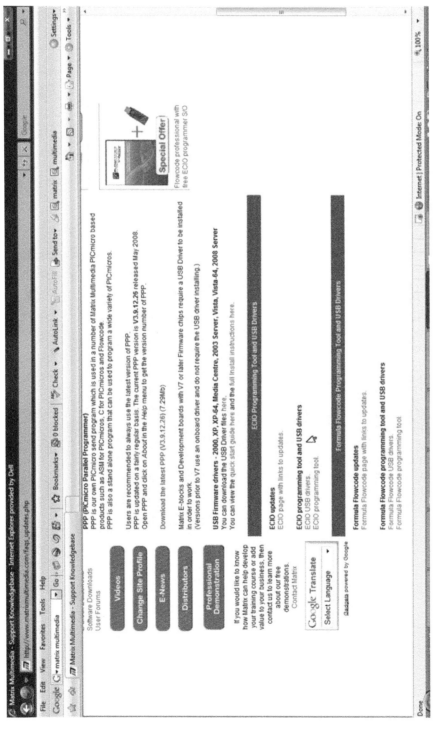

Figure A.1

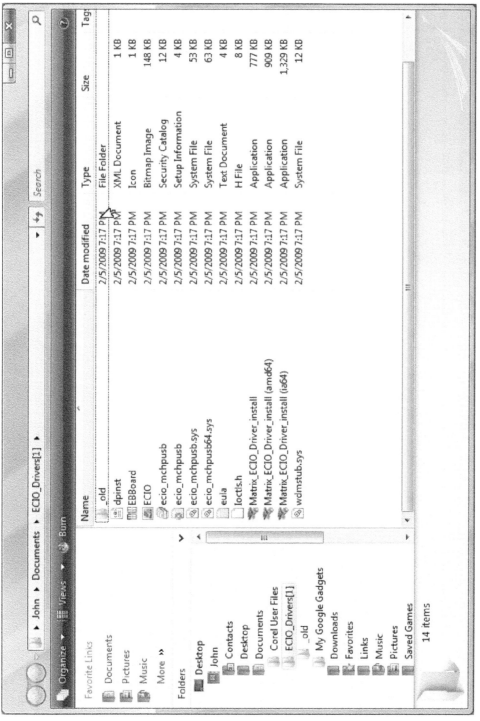

Figure A.2

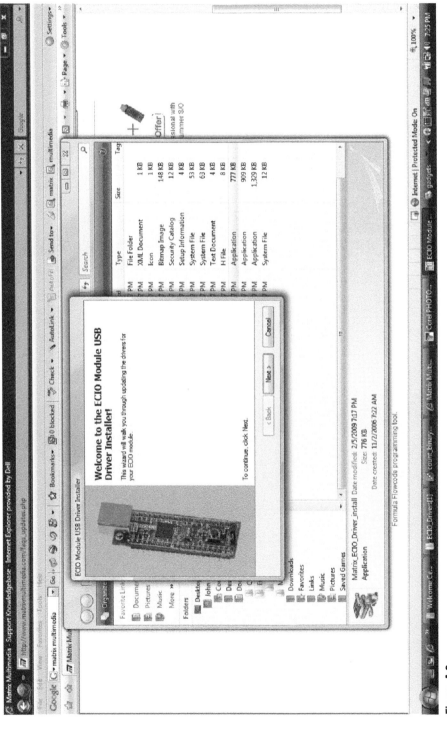

Figure A.3

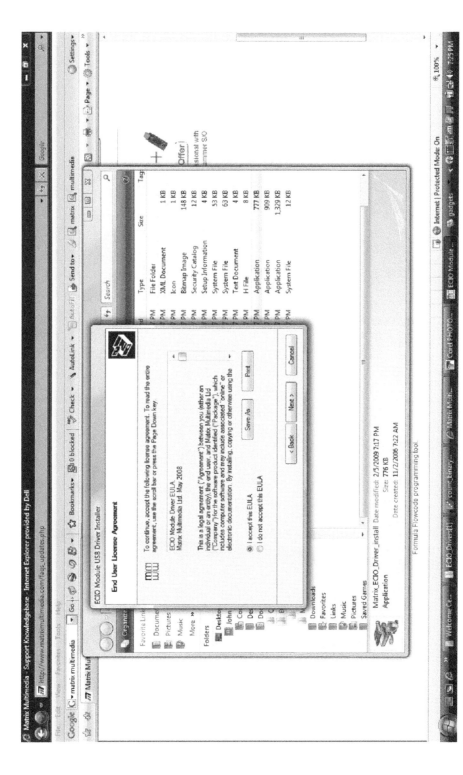

Figure A.4

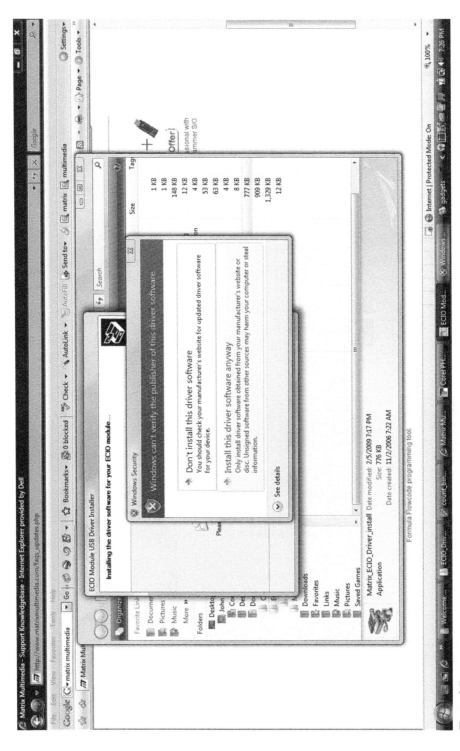

Figure A.5

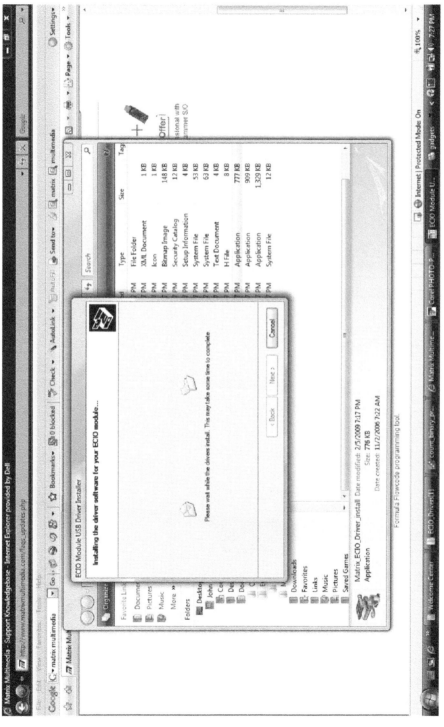

Figure A.6

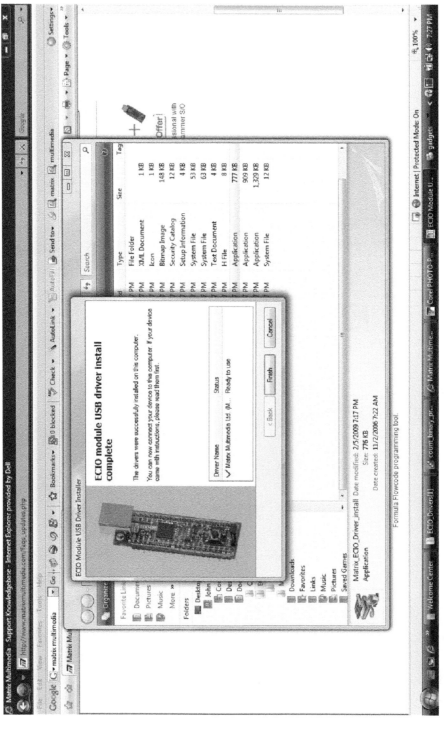

Figure A.7

INDEX

Printed and bound by CPI Group (UK) Ltd, Croydon, CR0 4YY

03/10/2024

01040341-0004